GEORGE JOHNSON'S
In the Palaces of Memory

"*In the Palaces of Memory* is a sophisticated, comprehensive and fascinating analysis of the science of memory. Johnson has achieved a rare blend of scientific and literary sophistication. Faithful to its complexities and controversies, the book is a fully dimensional portrait, a hologram, of the field."

—Richard Mark Friedhoff, *USA Today*

"As a writer about the biological and human sciences, Mr. Johnson has few peers. . . . If I wanted to give readers a feeling for the frontiers of cognitive neuroscience, I would send them directly to this book."

—Howard Gardner, *The New York Times*

"Among the liveliest and most readable accounts of neuroscience to have appeared in recent years . . . *In the Palaces of Memory* belongs in the library of everyone interested in the study of the brain."

—Timothy Ferris,
author of *Coming of the Age in the Milky Way*

"Lucid, insightful . . . Mr. Johnson imparts a huge amount of information in his clear descriptions of how the relevant experiments were conducted and interpreted. And he makes the reader feel the excitement that drives these people to devote most of their waking hours over a number of decades to the puzzle."

—John C. Marshall,
The New York Times Book Review

"An eloquent foray into how our brains convert experience into knowledge. [Written] with all the alacrity of a detective gathering clues . . . with a lucidity that at times approaches artistry."

—Thea Singer, *Boston Phoenix*

In the Palaces of Memory

GEORGE JOHNSON

In the Palaces of Memory

HOW WE BUILD THE WORLDS INSIDE OUR HEADS

VINTAGE BOOKS

A DIVISION OF RANDOM HOUSE, INC.

NEW YORK

FIRST VINTAGE BOOKS EDITION, MARCH 1992

Copyright © 1991 by George Johnson

Library of Congress Cataloging-in-Publication Data
Johnson, George, 1952 Jan. 20–
    In the palaces of memory : how we build the worlds inside our
heads / George Johnson.
        p. cm.
    Includes index.
    ISBN 0-679-73759-6
        1. Memory. 2. Neuropsychology. 3. Neurology. I. Title.
[QP406.J64 1992]
153.1'2—dc20   91-50556
                        CIP

Manufactured in the United States of America
10 9 8 7 6 5 4

For Ron Light

CONTENTS

PREFACE

Invisible Palaces

You can never dismantle all these modern mental structures. There are so many of them that they face you like an interminable vast city.

—SAUL BELLOW, *The Bellarosa Connection*

Whenever you read a book or have a conversation, the experience causes physical changes in your brain. In a matter of seconds, new circuits are formed, memories that can change forever the way you think about the world. I find that idea so remarkable that for the last three years it has been difficult for me to maintain much of an interest in anything else. How is it that memory leaves its mark so that we are able to carry around the past inside our heads?

It has only been in the last few years that scientists have begun to come up with good theories of how this might happen. By breaking down some of the walls that have traditionally divided their fields, psychologists, biologists, physicists, and philosophers are joining in an attempt to answer some of the great questions of human existence: How do we know what we know? How do we make the mental maps—the structures of memories—that serve as our guides to the world?

It's a little frightening to think that every time you walk away from an encounter, your brain has been altered, sometimes permanently. The obvious but disturbing truth is that people can impose these changes against your will. Someone can say something—an insult, a humiliation—and you carry it with you as long as you live. The memory is physically lodged inside you like a shard of glass healed inside a wound.

For someone like me who has always taken an absolutist view of the First Amendment, this idea raises problems that are difficult to resolve. Freedom of speech is based on the old dualist notion that mind and body are separate things. Hurting someone with a rock is different from hurting someone with an idea. But is it really? As science continues to make the case that memories cause physical changes, the distinction between mental violence, which is protected by law, and physical violence, which is illegal, is harder to understand. Who hasn't had the experience of seeing something so horrible and ugly that it was burned in the brain forever? I'll never forgive David Lynch for his movie *Eraserhead*.

. . .

Memory is clearly one of those subjects that can expand to engulf volumes. In an attempt to hold it at bay, I have written this book in three sections. Each is a story of someone—a biologist, a physicist, and a philosopher—who is trying to learn how memory works.

As in my last book, *Machinery of the Mind*, I'm writing about science in the making, not science already done. By working so close to the cutting edge, I've been able to watch the contest before the final results are in. Science is too often prettied up in retrospect. Once a theory becomes enshrined as the winner, the competition is quickly forgotten. But it's in the messy, sometimes irrational part of the scientific horse race that one can see a mind at work, confronting the unknown.

In most newspaper accounts, science comes off looking like butterfly-collecting. Facts are captured in a net, chloroformed, and pinned with labels on a Styrofoam board. The unruly, creative art of theory-building is rarely written about. If the subject is broached at all, the assumption is that the scientific "process" works as we were taught in high school: A scientist makes a hypothesis, then designs an experiment to test it. If the experiment confirms the prediction, the hypothesis is supported. If the experiment fails, the hypothesis is overturned.

But science hardly ever works that way. A scientist is no more ready to abandon an attractive idea than an author is to give up a novel in progress or a composer a symphony. Faced with an onslaught of conflicting evidence, a scientist will make minor adjustments in his theoretical edifice, moving a crossbar here, shoring up a support beam there, doing everything he can to keep his brainchild standing. At its worst, a troubled theory can become so laden with elaborations that it comes to resemble the conspiracy theories I wrote about in my first book, *Architects of Fear*—closed, hard-shelled systems of thought that resist any attempt to refute them.

Even when the scientific process works as it is supposed to, with reality-testing and competition weeding out the bad ideas, the question of how a theory matches up with some kind of real world is a lot more difficult than many scientists like to admit. Is a theory invented or discovered? Is science just another philosophy whose tenets include materialism and cause-and-effect? Do quarks and electrons exist in the way that marbles seem to, or are they convenient fictions, mental constructs that help us organize data in a useful way? To some degree the latter is true. But why then do our televisions work? Could we have a successful physics—one that resulted in nuclear power reactors and nuclear bombs—if we used other concepts than protons and neutrons, carving up the world differently? To what extent are the laws of science out in the world and to what extent are they inside our heads?

Somewhere toward the end of this project, I realized that it had come full circle. The question of what science is and why it is so successful comes back to memory and the way brains convert experience into knowledge.

. . .

In his book *The Memory Palace of Matteo Ricci*, Jonathan Spence writes about a sixteenth-century Jesuit who brought to the people of China a wonderful memory system that had been used in the West since the days of ancient Greece. To improve their powers of retention, people would build memory palaces, huge imaginary buildings they kept inside their heads. After years of practice, the images would become so vivid that a person could close his eyes and picture his palace as though it were real. Eventually, these mental architectures would become impossible to erase.

If an orator wanted to memorize a speech or a tax collector wanted to remember a list of names, he would mentally place each item inside a room in his own personal memory palace. When he wanted to recall the information, he would enter the front door and wander from room to room, retrieving the images. The palace was a structure for arranging knowledge.

"To everything that we wish to remember," wrote Ricci, "we

should give an image; and to every one of these images we should assign a position where it can repose peacefully until we are ready to claim it by an act of memory."

The art of building memory palaces has become a historical curiosity. But in a way, we are unconsciously building structures like this all the time. What Ricci taught as a deliberate mnemonic device comes close to describing what the brain does automatically. As we move through the world, experience is transformed into memories. Neuron by neuron, we snap together mental structures, constantly evolving palaces of memory that we carry with us until we die. Unlike Ricci's palaces, ours are invisible to introspection, but slowly scientists and philosophers are learning to read the blueprints of these worlds inside our heads.

In the Palaces of Memory

PRELUDE

The Tower in the Jungle

ABOUT FIVE YEARS AGO, Gary Lynch, a biologist whose specialty is the chemistry of human memory, found himself in a situation that seemed almost inconceivable to him. He was sitting on a stage at the University of California's Irvine campus in Orange County, California, describing his research to members of what he once would have considered warring tribes. He was surrounded by psychologists, linguists, philosophers—even a few computer scientists—all brought together for an annual meeting of a recently formed organization called the Cognitive Science Society.

As he gazed at the audience, he thought, What the hell am I doing here?

"I really felt like it was a big plain out there," Lynch recalled, "and all these different tribes were sending representatives and asking, 'How many people are in your tribe?' and 'What are your totems?' and 'What are your customs?' "

He half expected them to exchange beads and trinkets instead of conversation.

Before him that day were tiers upon tiers of seats filled with people who would seem to have a lot in common. They all shared an interest in this thing called memory. Yet historically, they haven't felt much in the way of kinship. They have looked upon one another as interlopers, not colleagues, jealously guarding their ancestral lands in the vast, unexplored jungles of the brain and mind.

In the last decade, some of these groups had forged uneasy alliances, a shaky coalition that was coming to be called cognitive science. Cases had been reported where psychologists actually conferred with philosophers, seeking clues for how to shore up their theories; some psychologists picked the brains of computer scientists for insights about how hardware and software might help them think about the connection between the brain and mind. It was even said that computer scientists would occasionally send ambassadors to the more ethereal disciplines, like linguistics, for ideas about how to design better programming languages. But much of the time, these groups still eyed one another with suspicion, agreeing, however, that neurobiologists like Gary Lynch were hardly worth talking to at all.

Standing at the edge of an intellectual wilderness, cognitive science was like a great teetering tower rising over the terrain. At the top (some would say the attic) were the philosophers, who looked at the mind from the most abstract and lofty level. Along with such rarefied notions as truth, beauty, the meaning of life, and the meaning of meaning, the philosophers reflected on the nature of intelligence. But only occasionally did they look down to the floors below them, where psychologists toiled at their own level of abstraction, designing experiments and gathering clues about the mind. The philosophers thought the psychologists lacked direction, that when they left the tower for one of their hunting expeditions they wandered blindly, bumping into things. For want of a guiding light, they couldn't see the forest for the trees. The philosophers, of course, never left the building, and the psychologists had little regard for speculations about forestry coming from people who didn't know a ponderosa pine from a Douglas fir. The linguists and the computer scientists occupied floors somewhere near the psychologists. All these groups hovered toward the top and middle tiers of the edifice, with the neurobiologists, like maintenance workers, stationed many floors below.

To those who preferred to deal with the mind from on high, the people who mucked around in the wetware—the three pounds of gray matter that fills the hole inside our heads—were indisputably outsiders, with a language that was indecipherable and a style of research they were welcome to call their own. Of course it was important for *someone* to understand how neurons worked. But the psychologists were convinced that the answers to the big questions—What is mind? What is consciousness? What is memory?—lay in the abstract realm of psychology; the philosophers believed they would find the answers in philosophy; the linguists in linguistics; the computer scientists in computational theory. There was very little sense that these fields could all be parts of a greater whole, much less that together they could benefit from a deeper understanding of calcium currents, potassium currents, sodium currents, serotonin, acetylcholine, norepinephrine—all the messy chemical complexities churning about in the basement of the mind. The inhabitants of the tower's upper regions had about as much interest in these details of neurobiology as an auto mechanic would in organic chemistry and geology, the sciences that explain how swamps and dinosaurs became motor oil. And the neurobiologists, who had actually held brain tissue in their fingers, looked upon these other explorers of the cognitive realm as a carpenter might look at a postmodern architect, or a novelist at a literary critic steeped in the arcana of the deconstructionist school—as dilettantes who had soared so high into the stratosphere of abstraction that they had lost touch with the matter at hand, who had become so mesmerized by their ideas that they traded them for the real thing.

That, anyway, was how the situation had often seemed. Recently, though, something was changing. Lynch was finding that his rivals in these other disciplines were becoming more interested in what he and his colleagues had to say. In the last few years biologists had been accumulating a staggering amount of information. Some of this raw material was beginning to congeal into theories. Lynch himself believed that he had identified the chemical process by which we convert experience into memory—into something solid and physical that can be lodged inside the brain. His findings were still too controversial to be considered monumental. No monolith had appeared on the veldt, shocking the apes into creating civilization. It would take more than one discovery to bring all these people together.

But during the next few years, as the 1980s became the 1990s, Lynch and other biologists would develop theories tracing memory to its very roots inside the neurons and synapses that make up the brain. At the same time, some psychologists would join with computer scientists to develop a hybrid called neural network theory, in which computers were programmed to model thousands of neurons working together, to mimic a piece of the brain. For centuries philosophers have debated what they call the epistemological question: How do we know what we know? As biologists studied memory at the level of the single synapse and the network people studied how hordes of neurons cooperated to form mental maps, they were engaging in what might be called applied philosophy. Even the physicists were getting involved, finding strange parallels between brains and other exceedingly complex systems.

"I think this is going to be one of the great playgrounds for intellectuals for the rest of this century," Lynch said. "I think it's possible that brain systems for learning and memory will become something like what Darwinian evolution provided in the late nineteenth century and early twentieth century—which is a playground. You had biochemists, political philosophers, economists, behavioral scientists, psychologists, psychoanalysts. You had people of every stripe and caliber, even amateur bone diggers—everybody could play this game."

Darwin provided a lens that brought together all these scattered beams. Now a theory of memory seemed in sight, one that would draw from biology, psychology, computer science, physics, and philosophy. The goal was to explain not only how we store individual facts but how we weave them together into a world view.

"In the end, all these groups of people are sort of hunkering down to talk about the same thing," Lynch said. "It's really quite an unprecedented moment to see this gathering of so many diverse tribes. It's as though some dark star appeared and it has an enormous gravitational force, and it's inevitably going to shape the future of all these fields in just the same way that Darwinian evolution did. For the first time I've discovered how much easier it is for me to pass knowledge between fields, which was a big problem. Now we have this common language."

PART ONE

Mucking Around in the Wetware

Nature is not mute. It eternally repeats the same notes which reach us from afar, muffled, with neither harmony nor melody. But we cannot do without melody. . . . It is up to us to strike the chords, to write the score, to bring forth the symphony, to give the sounds a form that, without us, they do not have.

—François Jacob, *The Statue Within*

A Dark
Continent

WHEN GARY LYNCH wants to instill in his students a sense of what it is like to do science—to wrest from the world's complexity a fact that is clear, simple, and indubitably true—he tells them the story of Emil Du Bois-Reymond. In 1843, in one of those triumphant moments that all scientists strive for and few ever attain, this little-known physiologist became the first person to see beyond doubt that electricity—and not some supernatural "life force"—runs through the nervous system. Working with an apparatus of electrodes and wires, he demonstrated the existence of what is now called the "action potential," the electrochemical pulse that beats in our neurons with a rhythm that is no less than the language of the brain.

"It is hard to believe," Lynch said, marveling at the intellectual climate in which Du Bois-Reymond's discovery was made. There were no telegraphs, no telephones, even Edison's light bulb was three decades away. Nobody yet knew that electricity consisted of electrons loosed from their atomic moorings and made to run through wires; that, cast into patterns, this flow could be used to carry information: the dots and dashes of Morse code, the sine waves that imitate sound and light, the binary logic that animates the chips of a digital computer.

With his kite and key, Benjamin Franklin had shown that electricity came from the sky as lightning and could be stored like an invisible fluid in a foil-wrapped vessel called a Leyden jar. In the 1790s Luigi Galvani showed that when applied to the severed legs of frogs, electricity made them twitch, as though momentarily rejuvenated. But it was one thing to show that a current can stimulate a dissected muscle and quite another to establish that inside a living body electricity actually travels the pathways of the nervous system.

"If I do not greatly deceive myself," Du Bois-Reymond had written, "I have succeeded in realizing the centuries-long dream of the physiologists—the equation of the life-force with electricity."

In all the history of science, Lynch says, that is his favorite quotation. "Imagine what it must have been like for him, when he sat there and thought, Oh, my God, it *really is* electricity going down that thing. *It really is electricity!* The same electricity that we know from a battery is going down that nerve!"

With his gapped teeth, curly hair, and mischievous eyes, Lynch sometimes looks like a grown-up version of Alfred E. Newman, right down to his "What, me worry?" smile. In more serious moments, he bears something of a resemblance to Bob Dylan. Chomping on a cigar, he speaks in long, spiraling monologues. He is a natural teacher, and he seems to enjoy nothing more than trying to describe the complexities of neurobiology in language an outsider can understand. For the last decade Lynch, a senior professor at the Center for the Neurobiology of Learning and Memory at the University of California's Irvine campus, has been trying to understand memory on the same level that we now understand digestion, respiration, and circulation—as something biological.

His tools are microelectrodes and scalpels, his subject matter neurons excised from the brains of rats. He floats this neural tissue in the sustaining fluids of petri dishes and measures the tiny voltages. He slices it into cross sections less than a micron thick so they can be photographed with the penetrating beam of an electron microscope. He is looking for the trace that is left when an event is recorded inside us. He is searching among the neurons for these elusive things called memories. And he is hoping for the same kind of experience Du Bois-Reymond had a century ago when he discovered the action potential.

So much of science has become maddeningly indirect. Scientists pile inference upon inference, building great logical towers; computers analyze reams of data for the statistical patterns that now pass as truth. This kind of analysis is an important part of Lynch's work, but sometimes he finds it unsatisfying, too many levels removed from direct experience, from that rare, existential moment when, by the very act of observation, subject and object fuse in a jolt of recognition, when we crystallize from the haze of potentiality a fact.

"There is no substitute for that experience," Lynch said. "And it is an experience that remarkably few people ever have. You'd think that it is part and parcel of the scientific process, but it's not. Once you've had the feeling, it's like trying to tell people what it's like to see color when they don't see color. There's no real explaining it. It's not even a level of intellectual understanding; it's a level of emotional understanding almost. It's a sort of satisfaction that you really *have* the thing."

Since the early 1980s, Lynch has been trying to identify a very specific biochemical reaction that he believes forms the infrastructure of memory. If he is right, the breaking of a single kind of molecule inside the neurons explains how experience causes the brain to change.

"We are doing what people once thought was impossible," said Lynch, who is not known for his modesty. "We are watching the formation of a specific piece of memory. I no longer find it inconceivable that we could in the future be able to think about how we form concepts, why we have such vast memory capacity, how we retain sequentiality and spatial information in our mental maps, and even how we funnel all the information from one region of the brain to the other as we go through cognitive-like steps. Those things are no longer so mysterious to me as to be impenetrable.

"That's not to say that my ideas are right; it's to say that *something* like this is going to be it."

Lynch's findings are controversial and far from complete. But in the cautious realm of neurobiology, where careers are spent mapping a small part of the brain's neural confusion or studying the chemistry of a single neurotransmitter, his hypothesis is refreshing for its boldness and scope. It is unusual to see someone attempt so grand a synthesis, to cut through the ambiguity and uncertainty and say *this* is the way memory works.

"Wandering around on a dark continent" is how Lynch once described his quest. But occasionally, one stumbles upon an unexpected vista and can suddenly see for miles.

Looking
for Engrams

IN 1950 KARL LASHLEY, one of the most prominent neurological researchers of his day, wrote an influential paper called "In Search of
the Engram," in which he looked back on decades of failed efforts to
discover where memories reside. When we listen to a symphony or
jazz, a melody is somehow impressed within us. And we recognize
it when we encounter it again, not only during that evening's performance when, after half an hour or so of wandering, the orchestra
or soloist returns to the original theme, but the next time we hear the
piece—a day later, a week, several years or decades. We hear a tune
for several seconds and it leaves a trace—an engram, Lashley called
it—that lasts until we die.

How can something as evanescent as a memory take on substance
and become part of the brain, part of the body? Centuries ago the
British empiricists suggested that information flowed in through the
senses and was impressed on the brain as though it were a clay tablet.
Each memory left a marking, an engram. While the empiricists thought
we were born with the tablet empty, a blank slate, Immanuel Kant
believed that we entered life already equipped with some of the knowledge necessary for interpreting the world outside. But as for the details, Kant was as sketchy as the empiricists. How was this information
stored, that which was innate and that which was acquired? Obviously, we don't have little words and pictures inside our heads. But
what is the internal language in which the stuff of life is written? The
image of the clay tablet almost suggests some kind of cuneiform,
patterns of sharp little gouges recording everything we know.

The technologies of the late-nineteenth and early-twentieth centuries suggested more likely metaphors. Throughout history, messages have been sent through space by translating them into some
kind of code. Information can be transmitted across a distance, provided that the sender and receiver share a common set of symbols:

one if by land, two if by sea; three puffs of smoke means trouble ahead. Samuel Morse showed just how precisely this mechanism could be honed when he invented the telegraph, allowing whole texts to be transmitted for miles using dots and dashes of electricity. Morse's signals traveled through wires; then Hertz and Marconi showed how messages could be broadcast through space using electromagnetic waves. With Edison's phonograph, sounds could be saved as squiggles on a foil-wrapped cylinder and played time and again.

With the invention of radio, television, and tape recording, it became clear that both sounds and pictures could be transmitted and stored using patterns of electromagnetism. By the time it was discovered that electromagnetic waves emanate from the brain, the idea of storing information in some physical medium—be it a spool of tape or a glob of neurons—was slightly less mysterious. Still, in the case of the brain, no one had any idea how the recording was done. In attempting to explain how memory works, Lashley and his colleagues might have felt that they were not much better off than Aristotle, who assumed that the mind was centered in the heart, not the head.

. . .

Lashley had approached the question in a manner typical of his day. In a series of experiments beginning in the 1920s, he trained rats to run a maze. Then, after cutting out a tiny bit of an animal's brain, he would set it loose in the maze again. It would seem that when, by chance, he had snipped away the bit of tissue containing the map of the labyrinth, the rat would suddenly forget what it had learned— the engram would be gone. With one slice of the scalpel, what was familiar would become strange.

But after hacking away at the brains of a number of rats, Lashley was never able to find a single location where the memory was recorded. As he destroyed more and more of the animal's brain, it would become increasingly sluggish and less adept at navigating the corridors. The less brain a rat has, the worse it is at running mazes— nothing surprising about that. But what puzzled Lashley was that it didn't seem to matter what part of the brain he eliminated. As the volume of the brain was gradually reduced, the memory of the maze degraded, but no single snip of tissue would make it disappear. The engram didn't seem to exist.

"This series of experiments . . . has discovered nothing directly of the real nature of the engram," Lashley ruefully concluded. "I sometimes feel, in reviewing the evidence on the localization of the memory trace, that the necessary conclusion is that learning just is not possible."

Of course he was writing tongue in cheek. What Lashley had decided was that a memory does not exist in any single place, like a folder in a file cabinet, but is somehow spread like smoke throughout the brain. For those who believed that the mind was a mysterious substance separate from the brain—the ghost in the machine—Lashley's holistic theory provided some strong theoretical ammunition. Memories indeed seemed like ghosts. Most scientists kept insisting that the brain was some kind of very complicated biological machine. But what kind of physical device could act in a way that meshed with Lashley's experiments?

Beginning in the 1950s, a few neuroscientists seized on a new metaphor, one that suggested a physical explanation for how a memory might permeate large regions of the brain. Using laser beams, scientists had learned how to make an eerie kind of three-dimensional photograph called a hologram. Viewed with the proper illumination, the image stored in a hologram seemed as solid as the little piece of reality that had been recorded. It was striking enough that a two-dimensional piece of film could be used to store and project a three-dimensional image. Stranger still, when a hologram was cut into pieces, each fragment retained the entire image, though with poorer resolution. Was it possible that the brain was like a hologram, with each tiny piece of neural tissue containing everything an animal knew?

Most scientists found the evidence for holographic neurons dubious at best. In direct competition with Lashley's holistic school were the localizationists, who continued to hold that memories were located in specific places in the brain. At about the same time that Lashley was lobotomizing rats, a Canadian surgeon named Wilder Penfield was uncovering a very different story. During a series of "open brain" operations, Penfield stumbled upon dramatic evidence that engrams existed—and that they could be selected and played like records in a jukebox.

Penfield worked with epileptics. By opening a patient's skull and probing the surface of the brain with an electrode, he hoped to find the region from which the seizures emanated—the epicenter of the quakes. During the operation, it was necessary to keep the patient conscious. Penfield found to his surprise that when he touched his electrode in one place, a patient would think he had heard a sound; touch another spot, and the patient would see a flash of light. Some locations seemed to hold memories of melodies or incidents from childhood. At the touch of an electrode, one woman felt that she was in her kitchen, listening to her boy playing outside; she worried at the sound of passing cars. A young man relived the experience of sitting at a baseball game, watching a child crawl under the fence to sneak inside. Each time Penfield stimulated the spot, the memory would be played again.

"The astonishing aspect of the phenomenon," he later wrote, "is that suddenly [the patient] is aware of all that was in his mind during an earlier strip of time. It is the stream of a former consciousness flowing again. If music is heard, it may be an orchestra or voice or piano. Sometimes he is aware of all he was seeing at the moment; sometimes he is aware only of the music. It stops when the electrode is lifted. It may be repeated (even many times) if the electrode is replaced without too long a delay."

Some of his colleagues wondered if Penfield was really tapping into memories. The recollections the patients described sometimes sounded more like hallucinations. Even if these were real experiences that were being replayed, nothing explained how they were recorded in the biological medium of the brain. While Lashley's work on memory suggested the metaphor of laser holography, Penfield's suggested something like a video recorder. But neither model was very convincing. Each obscured more than it explained.

· · ·

As the digital computer rose to power in the second half of the twentieth century, the localizationist view became dominant. In a computer, memories are stored in very precise locations. Why should it be different in the brain? A number of psychologists were seized by this idea that the mind could be thought of as software running on some sort of biological machine. "The mind is what the brain does" became their battle cry. While this was a neat way to argue against dualism—the idea that the brain is inhabited by a separate, ethereal mind stuff—the biologists were not very impressed. When it came to memory, the computer metaphor was not much more illuminating than its predecessors. After all, a computer doesn't really remember any more than a video camera sees. In a computer, what passes for memory consists of the 1s and 0s of binary code stored in a bank of transistors, the precursor of the chip, or on a spinning magnetic drum. The computer metaphor was just a fancier version of the video recorder model. Maybe on some level the brain was a kind of computing machine. But nothing explained how it could store such a vast amount of information, not simply recording it but actively arranging and rearranging it into structures, fitting in a new memory among everything else that was already known.

While the computer model of the mind continued to enchant the psychologists, the search for the engram moved to different ground. Inspired by Watson and Crick's discovery of the double helical structure of DNA, a few biologists began to consider an entirely different storage site, the molecules inside the brain. If a sequence of molecules called nucleotides—the steps on the helical staircase—could encode the genetic information necessary to make a human, why couldn't memories be recorded this way? The alphabet of memory would be the letters A, C, T, and G—the molecules adenine, cytosine, thymine, and guanine that spell the instructions for making enzymes and other proteins, the very substance of life. While it was not at all clear how this four-letter code would spell out a memory, much less a whole childhood experience, the notion of a biological code whose symbols were molecules was hard to resist. How wonderful it would be if evolution had taken the same mechanism used to store a species' genetic memory and adapted it for use in the brain.

For a while it seemed that this might be the metaphor the neuroscientists were seeking. In 1965 a neurobiologist named Allan Jacobson reported that he had trained rats to react to a flashing light by heading for their food dispensers, where they were rewarded with nourishment. Jacobson killed the animals, extracted RNA (an information-carrying molecule similar to DNA) from their brains, and injected it into the stomachs of untrained rats. Then he would test these animals by flashing a light and seeing how they reacted. Sure enough, Jacobson claimed, the rats would tend to head for the food dispenser, as though they had gone through the training sequence. A memory, it seemed, had been taken from the brain of one rat and squirted into another. The engram appeared to be something that could be carried around in a syringe. In other experiments, worms called Planaria were trained to avoid light, then chopped up and fed to other Planaria, which seemed to inherit the trait.

One researcher, Georges Ungar, insisted that memory was encoded not in nucleic acids but in a different molecular alphabet: the sequence of amino acids that make up protein chains. Working in the early 1970s, he used electrical shocks to train rats to avoid darkness. Then he extracted chemicals from their brains. By analyzing this mixture, he found a proteinlike substance (a polypeptide consisting of eight to fifteen amino acids) that seemed to contain the memory of the electrical shock.

Other rats injected with this molecule, or even a synthesized version of it, also tended to avoid the dark. Ungar called the chemical scotophobin—derived from the Greek words for "fear of the dark"— and claimed to have found similar molecules that carried other memories. A few people imagined the day when pills would replace books. Starving M.B.A. students could sell brain fluid to pharmaceutical companies instead of blood plasma. But most researchers remained skeptical. In all these cases, the evidence was statistical, unconvincing, and impossible to replicate. With Ungar's death in 1977, research into chemical engrams lapsed into obscurity. Now most neuroscientists believe that scotophobin is to psychology what phlogiston is to chemistry—a figment of the imagination.

. . .

Today these molecular theories of memory are regarded as violations of what has become a central dogma of neuroscience: Memory exists not at the level of the molecules that make up the neurons but at the level of the neurons that make up the brain. Proteins, after all, have very brief lives, far shorter than that of most memories. For each new engram, the brain would not only have to make a new kind of protein; it would also have to develop a mechanism for constantly replenishing the supply. Neurons—and, more important, patterns of neurons—last far longer, generally for a lifetime. Brain cells, not molecules, seem better equipped to act as the biochemical letters in the mental archives.

According to the dominant view of neuroscience that has developed over the last few decades, the brain is indeed a kind of computer, but only in the most general sense. In fact it sometimes seems more revealing to think of a single neuron as a little computer and the brain as a network of tens of billions of these information-processing cells. Each neuron receives electrical impulses through a treelike structure called a dendrite, whose thousands of tiny branches funnel signals into the body of the cell. In computer jargon, the dendrite is the neuron's input device. While some of the arriving signals stimulate the neuron, others inhibit it. If the pluses exceed the minuses, the neuron fires, sending its own pulse down a stalk called an axon. The axon is the output channel. It feeds, through junctions called synapses, into the dendrites of other cells. The resulting circuitry is complex beyond imagination. A single neuron can receive signals from thousands of other neurons; its axon can branch repeatedly, sending signals to thousands more.

As important as the neuron itself are the synapses that serve as junctions between the cells. While information is carried inside a neuron by electrical pulses, once the signal reaches the end of the axon it must be ferried across the synaptic gap by chemicals called neurotransmitters. On the other side of the synapse, the dendrite contains structures called receptors, which recognize these transmitting molecules. If enough are registered, then the second cell fires. A neuron can be thought of as a cell whose specialty is converting chemical signals to electrical signals, then back to chemical signals again.

Scientists have known about the basic operation of neurons for decades. But it has only been in recent years that they have been able to explain how these information processors might be used to store memories. While the details are disputed, to say the least, the general idea is this: When the brain is exposed to a new event (the image of a face, the sound of laughter), a unique pattern of neurons is activated somehow. Within the vast web of brain cells, a constellation lights up. Unless this configuration is to fade with whatever evoked it, there must be a means of preserving it—of forging connections between

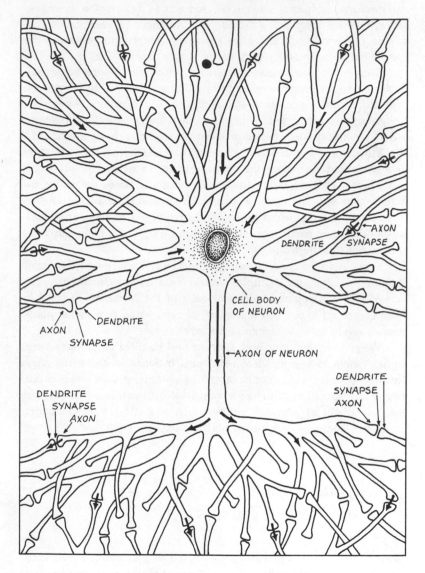

AXON
SYNAPSE
DENDRITE
CELL BODY
OF NEURON
AXON
DENDRITE
SYNAPSE
AXON OF NEURON
DENDRITE
SYNAPSE
AXON
DENDRITE
SYNAPSE
AXON

A NEURON AND ITS TANGLE OF CONNECTIONS.
*Signals are funneled into the cell through the dendrite and sent out
through the axon.*

the neurons, creating a new circuit that acts as a symbol, a representation of something in the outside world. Then, by reactivating the circuit, the brain can retrieve the memory—rough and perhaps a bit faded, but a serviceable replica of the original perception.

Recognition would occur when we encountered something that evoked a neural pattern similar to one that was already in storage. A reproduction of a painting in a book might cause a pattern of neurons to light up that resembled the pattern which had been welded together when we had the experience of seeing the original hanging in, say, the Museum of Modern Art. The brain would detect the similarity and we would experience the pleasant shock of recognition.

These patterns would not be so fixed and rigid as the wiring in a television set. The circuitry would be malleable, changing all the time. As knowledge was acquired, old circuits would break apart and form new connections, constantly building and rebuilding our representations of the world.

Ideas as well as specific images might also be acquired this way. A concept—*chair*, for example—would be a neural circuit that preserved the overlapping features of each of the hundreds of specific chairs recorded in our vast repository of experience. In a similar manner we might learn to recognize an artist's style.

We can see then how both Lashley and Penfield might have been right. A memory as complex as an evening in Santa Fe, or the trajectory through a maze, would not be stored in a single, precise location but rather as a sprawling structure of neurons. And yet, like a circuit, the memory might be activated by the touch of a stimulating electrode, as though someone had turned on a switch.

The Brain's Volume Controls

WHILE THE IDEA that memories are stored as patterns of newly connected neurons has only recently dominated the field, it was first suggested in 1893 by the Spanish neuroanatomist Santiago Ramón y Cajal. Half a century later, in his book *The Organization of Behavior*, the Canadian psychologist Donald Hebb revived the idea by speculating in some detail about how the new circuitry might be made: If, during a learning experience, one neuron tends to fire another, then the synapse between them will somehow be strengthened, linking them more firmly into the same circuit. Some of Hebb's disciples went on to suggest a more general version of the rule: Two neurons that tend to be active at the same time will automatically form a new connection. If they are already weakly connected, the synapse between them will be strengthened; if not, an entirely new synapse will be made. Hebb had no idea how any of this might happen. But, taking his hypothetical mechanism—which has come to be called the Hebb synapse—as a building block, he erected an elaborate learning theory in which neurons are connected to form neural assemblies, which are integrated into still larger structures called phase sequences.

The nomenclature of Hebb's theory has fallen by the wayside, but he provided a guiding light: Constellations of neurons can be used as symbols that stand for the things and ideas that make up our world. If a memory is a pattern, or circuit, of "lit up" neurons, like letters or pictures on an electronic scoreboard, then Hebb's rule provided a means for stringing them together, light by light. *Two neurons that fire together will form a link.*

It would be decades before biologists refined this notion, learning something about how the connections are actually made. This idea for building new brain circuitry has become so basic to the new theories of memory that it is easy to forget that when Hebb came up with it, there was no evidence for it whatsoever. In fact, there was no reason to believe that experience caused any kind of physiological change. For all that scientists could say, the dualists might be right after all: Memories really were apparitions with no physical basis in brain tissue. Someone would have to show that a brain that had learned something was physically different from the way it was before. With humans this kind of demonstration was nowhere in sight. But, beginning in the 1960s, Eric Kandel of Columbia University did what might be regarded as the next best thing. He showed how a crude form of learning causes changes in the simple nervous system of a sea snail called Aplysia.

From the standpoint of the neurobiologist, the Aplysia is a refreshingly uncomplicated creature. Its nervous system is "built like an old Philco radio," as one researcher has described it, "with simple circuits and large, easily identifiable components." Kandel decided he would open up the back of this biological appliance and probe among its neurological tubes, resistors, and capacitors until he understood how the system worked—and how it became modified by experience. He began by teaching the Aplysia a few simple tricks. Then he would study the animal's neural circuitry to see if learning had made it change.

Tap on the Aplysia's siphon, or spout, Kandel discovered, and it would automatically withdraw its gill. But if the tapping continued, the reflex would slowly weaken. The animal would begin to disregard the stimulus. This rather unimpressive phenomenon, called habituation, is comparable to our ability to adapt to traffic noise or the repetitious sounds of a Philip Glass composition. Still, it can be considered a primitive kind of learning: An organism modifies its behavior in response to a signal from the outside world.

The opposite of habituation is sensitization, in which a creature's nervous system becomes more receptive to a stimulus. Kandel produced sensitization in the Aplysia by giving an electrical shock to its tail. This jolt seemed to stimulate the snail's nervous system, making it hypersensitive. Now when its siphon was tapped, it overreacted, withdrawing its gill at the slightest touch. In both of these cases, habituation and sensitization, the Aplysia's nervous system had become readjusted somehow.

If the human brain can be thought of as a complex biological computer, the Aplysia's gill-withdrawal mechanism is more like a simple circuit consisting of a light bulb and a dimmer switch. Kandel found that the synapse seemed to function like a volume control. Or, perhaps more appropriately, a water faucet. In the case of habituation, he discovered, neurons in the snail's gill-withdrawal circuit were altered by the repeated tapping so that they released less neurotransmitter into the synaptic gap. Thus they sent weaker signals to the next neurons in the chain. In the case of sensitization, neurons were changed so that they released more transmitter.

Simple as it seems, this notion of synapses as volume controls has turned out to be a powerful concept. If it weren't for the buffer provided by the synapses, a signal injected into one neuron in the brain's dense web would spread like fire in every direction, resulting in a massive cerebral short circuit. By acting as volume controls, the synapses channel the electrical flow into useful directions. If a synapse is turned up, the neurons on either side become more strongly connected: If the first one fires, the next is likely to follow. On the other hand, two neurons whose synapse is turned down are, in effect, disconnected. They might as well be on opposite sides of the brain. The synapses, then, seem to allow for the malleability needed for learning. By adjusting the threshold of trillions of synapses, the brain can organize itself from an amorphous tangle of neurons into a complex information processor. Within the infinite chaos of potential connections, distinct circuitry is carved.

Although Kandel continues to concentrate on the simple reflexive behavior of invertebrates, he believes he is uncovering basic mechanisms that, like DNA replication and protein synthesis, could be fundamental to humans as well as snails. He likes to describe his work as a search for "the cellular alphabet of memory." But, as one neurobiologist put it, "it may not have all the letters, and that means you just cannot make certain words. Let's face it, this animal lies in the shallows and gets buffeted about by the water, and its normal ecology really doesn't require too much strain on its memory."

Gary Lynch belongs to a school of researchers studying synaptic change in man's closer relations: monkeys, rabbits, and rats. While some of these researchers are primarily concerned with anatomy— where in the brain different kinds of memories are stored—others, like Lynch, are looking deeper, at the biochemical level, where the most fundamental answers lie. They hope to move beyond such rudimentary forms of learning as sensitization and habituation and see how the brain stores complex memories, snapping together, neuron by neuron, the structures that serve as maps of the world.

Early
Obsessions

LYNCH BECAME INVOLVED in the search for the engram in a rather roundabout way. In the early 1960s, unsure whether he wanted to pursue English or electrical engineering (he had heard from a high school classmate that the money was better there), he enrolled at the University of Delaware. Though he had made good grades in high school and had assumed he would go to college, he found he had little patience for things academic.

"I hated school, and I hated anything associated with it," he recalled. "I hated the teachers. I hated anything remotely resembling authority, and I certainly found the classes boring and tedious and horrible.

"I've always had the kind of personality, that's—I don't know—*obsessive.*"

He had, in short, an attitude problem. He quickly became bored—especially with the science classes—and was expelled from school in his sophomore year "because of a lot of drinking and things like that." About six months later, he was allowed to return. He was grateful for the second chance, but convinced that he had no interest whatsoever in the hard sciences.

"I decided I was never in my life going to subject myself to anything as boring as biology or chemistry or physics again. I just became an 'undeclared' and studied mainly history and English and psychology."

The turning point came when he learned that one of his favorite professors was doing experiments that involved recording electrical signals in the brain. He encouraged Lynch to work on a project of his own.

"I enjoyed it. I couldn't stand the actual running of the experiments or setting up the experiments. That was horrible. But thinking about them was fun. We were having a ball. We were doing great things like seeing if we could detect ESP by brain waves. We would take a cat and hook it up with brain electrodes, and then down the hall we'd scare the hell out of its sibling. I mean, that's what kind of stuff you do when you're in college."

In more serious moments he began reading the literature on animal psychology, a century-long accrual of charts, formulas, and statistics seeking to quantify that most elusive of phenomena, complex behavior. In fact, he found he could spend hour after hour alone in the library, immersed in this vast body of knowledge, a world of facts that he felt compelled to explore.

"I've always enjoyed obscure forms of human knowledge," he said. "I don't know how to explain it. One of my hobbies is to read about the search for the historic Jesus: a lost little world of knowledge and a few linguists poring over old texts—that's the kind of thing. If I were doing it all over, probably what I'd get into is linguistics, and not neurolinguistics but textual criticism. Great stuff. There are little puzzles and little problems, and you don't have the outside world nagging you to death about what it means."

This is how he saw animal psychology, as a pure intellectual exercise, a world in and of itself whose complexities could be mastered in time.

"It had this marvelous, musty, esoteric feeling to it," he said. "I mean, its lack of utility was so evident to me; its lack of connection with the everyday world was so transparently obvious. You sit in a biology course and you have to listen to all this horrible stuff about eggs and sperm, and it's so immediate and proximal—it's the difference between studying current events and studying the medieval period. The medieval period is so much nicer. It's *there*, and you can put it down and come back and it will still be there, and you read about it. It's enjoyable and esoteric and kind of exciting. It exists unto itself, and there is only a small group of people who are really interested. That's how I found animal psychology, compared to biology. Animal psychology for me was dusty books in the back of the library. Biology was this stuff that—I never finished my course in biology when I was at the university. That's why it's marvelous to be a professor of biology because I never had a course in biology. I started one, and it was just too much."

After graduating from the University of Delaware in 1965, Lynch was accepted into Princeton as a graduate student in psychology.

"I really didn't see the point of a Ph.D.," he said, "but I observed that the professors at the university had what to me was the ideal life. I just loved the life-style they lived. I'm someone who came from a lower-middle-class environment. When I was out of school, in the summers, I worked in factories, and when I went through college I worked in all-night gas stations. That's work. Now I watched these guys at the university. They spent a lot of time at the tavern. They spent a lot of time chasing undergraduate and graduate students. They seemed to have an unbelievable amount of leisure time and no great stress in their occupation. And they were being paid, by my standards, very well for this. And believe me, this appealed to me immensely as opposed to a business job.

"And the other thing they had was a great tolerance. They liked me. I got along well with those people, which is certainly not the case in most situations I've been in in my life. I've always found that I had trouble with people. Authority."

As Lynch was becoming entranced with the idea of life as a professor he was also becoming seriously interested in the way the brain generates the mind. The idea that he could be paid to explore questions like that seemed almost too good to be true.

"I was beginning to realize that I was probably good at this," he said. "See, one of the major problems about education is that people at places like Delaware always have the idea that I'm good here, but if I was at Harvard I wouldn't be good. You find someone who is a very good undergraduate student and you wonder why they aren't better, and you find out it's really because they think somewhere around the corner are all these smart people and they're not one of them."

Lynch was beginning to feel that he was one of the smart people, that he was just as likely as anyone to solve some of the great problems of psychology.

"One of the things I discovered in graduate school was that I had a talent for writing papers, and a talent for poring over data. If I have any strengths as a scientist, one of them is this: I love to pore over the data. There's the same musty quality about it. And I just sit there and analyze it this way and that. Graph it this way, graph it that way. I can sit with data for hours, just as happy as can be. That's the secret of doing science. You can't imagine how profoundly important it is to sit and go through that material, and go through it so that you really understand what happens in the experiment—to gain it at an intuitive level."

. . .

What is it that allows scientists and other creatures to narrow their consciousness into so fine a beam and focus it on a single problem? And how are they able to keep from concentrating too intently, to the exclusion of everything else? While working on his doctorate at Princeton, Lynch decided to study this question of how animals control their level of attention.

While studying the anatomy of rats, he uncovered a neural circuit that went from the brain stem, where the spinal cord is connected, to the cerebral cortex—the part of the brain where we plan and make decisions—and then back to the brain stem. Lynch believed this was what electrical engineers called a negative feedback loop. The brain stem was known to control an animal's level of arousal, how awake it was. Lynch believed that as an animal started concentrating on a task—say, finding food in a maze—some of the increased nervous energy was fed back from the cortex to the brain stem. These signals would, in turn, elicit restraining signals that kept the animal from becoming so aroused by the task before it that it became hyperactive, spinning wildly out of control.

To test his theory, he agitated rats by depriving them of food or by giving them small doses of amphetamines. In normal rats, the regulation circuitry seemed to act as a flywheel, dampening the effect. But if he first "lesioned" the circuit by burning it with electricity, a rat would lose control and become too hyperactive to concentrate on much of anything. The control loop had become disconnected.

Thus, in a time-honored method of neuroscience, Lynch demonstrated the existence of a circuit by showing what happened when it was destroyed. He had made his first scientific discovery. With further experiments, he wrote up the results and speculated on the implications for learning and memory. After all, to learn one must be able to focus and pay attention. The higher brain, where consciousness seems to lie, must somehow communicate with the parts of the lower brain that control arousal.

"I really thought I was cooking," he said. "I published this paper in the *Journal of Comparative and Physiological Psychology* and confidently leaned back and expected the accolades of a startled world as these experiments popped on them. And I got *seven* reprint requests. You want to talk about depression! I intended to have a more rapid meteoric climb in the profession."

But then came an accidental discovery that indirectly led Lynch to the field of memory research. Three weeks after lesioning some of the rats, he gave them amphetamines again and found to his surprise that they no longer became uncontrollably excited. The circuit he had destroyed apparently had been repaired or replaced.

Lynch hadn't uncovered anything new. In the jargon of neuroscience, this phenomenon was called "recovery of function" and had been known about since the nineteenth century. Generally, brain damage lasts forever. Unlike skin cells, neurons are not replaced when they die. But if the *connections* between nerve cells—the dendrites, synapses, and axons—are damaged, they sometimes grow back.

By documenting an example of recovery of function, Lynch simply added a footnote to the literature; the underlying mechanism remained obscure. The discovery was less important to the history of neuroscience than to Lynch's growing self-confidence. He was still several years away from the beginning of his search for the chemistry of memory, but in retrospect he considers this the first step. He had seen for himself that the brain is not static. It can change—an obvious requirement if it is to store memories.

Growing
New Wires

AFTER RECEIVING his Ph.D. from Princeton in 1968, Lynch went to the University of California's Irvine campus, an island of higher learning in Orange County's suburban sprawl. In those days, Lynch had long hair, a beard, and an irreverent manner that came with the age. "I've always assumed that older people are not as smart as younger people," he often says. This attitude marked him as a sympathizer with the counterculture, a professional disadvantage in a county that was still a stronghold of the John Birch Society. But Lynch was popular with his students, winning an award for best teacher of the year. One alumna remembers going to Lynch as an undergraduate, seeking career guidance. Reaching in his desk for a flask of Jack Daniels, he offered her a drink. He was determined not to become an authority figure. Like a free university, his lab became known as a place where any student interested in how the brain worked could have a crack at the field.

To supplement the nine hundred dollars the school had given him to set up a laboratory, he applied to the National Institute of Mental Health for a grant to continue his work on the regulation of arousal. The professor he was working under had told him that members of the department were expected to file two grant applications. Not realizing that he was simply supposed to submit the same proposal to another agency, Lynch worked late into the night writing up a second proposal to study recovery of function and sent that to the National Science Foundation. The NIMH rejected his arousal request, ending forever that avenue of his research, but the people at the NSF responded enthusiastically. Understanding recovery of function, they hoped, would lead to a way of curing brain damage.

Lynch lost little time spending the grant money. Seeking a model for how the nervous system might repair itself, he turned to a badly understood phenomenon called sprouting—the growing of new neural connections. It had already been shown that sprouting occurs in neuromuscular junctions. When nerves are pulled from the muscles they stimulate, sometimes the connections grow back. New neurons are not generated, of course, but the damaged ones grow new dendrites, axons, and the synapses between them. More recently, sprouting had been shown to take place in the spinal cord. In 1969 a researcher named Geoffrey Raisman surprised the neuroscientific community by showing that sprouting could occur in the brain itself, in a structure called the septum.

"Most of us, I think, suspected that sprouting was something that was present in the neuromuscular junction and muscles, less present in the spinal cord, weakly present in the brain stem, still found in the septum—which is an older structure—but when we got into the cortex, or higher brain, we expected to see it disappear," Lynch said. After all, it was almost dogma that the adult brain does not change. Yet that presented a profound paradox. If the brain is indeed static, then how do we learn?

As far as Lynch was concerned, the septum barely qualified as brain tissue. It is part of the primitive lower brain that is involved in regulating the body's involuntary actions: breathing, digestion, maintaining blood pressure, and registering thirst, desire, or hunger. Lynch wanted to do Raisman one better and see if sprouting could take place in the higher regions of the brain—the cortex, where thinking occurs. The most advanced area of the cortex—the so-called neocortex, which is believed to be the seat of reasoning and consciousness—was hopelessly complex, providing too difficult a target. Lynch chose to concentrate instead on a simpler part called the hippocampus. Unlike other regions, the hippocampus was fairly well understood. Over the years it had been mapped with some precision, revealing that it is arranged in an unusually structured manner. While the septum is a tangle of connections, an anatomical mess, the hippocampus is neatly arranged in layers. If sprouting could be observed in this orderly array, it would be that much easier to tell in some detail just what was going on.

At the time Lynch was planning this new experiment, Walle Nauta, a professor at the Massachusetts Institute of Technology and the dean of American neuroanatomists, was visiting Irvine. Lynch idolized the man.

"He's very gentle, scholarly, and in every way a fine person," Lynch said. "Having read everything that was ever written by the guy, I considered Nauta the nearest thing to a god that I knew of. He was a colossus—was and is, a great, great scientist."

Lynch told him he was planning to attempt a sprouting experiment.

"Oh, what are you going to do?" Nauta asked. "That's an interesting effect, that sprouting."

Lynch said he was going to do sprouting in the hippocampus.

"Yes, that's a great choice, the hippocampus," Nauta replied. "That's a beautiful model. What is your experiment?"

Lynch carefully began to explain. Of the major inputs connecting the hippocampus to other parts of the brain, one comes from a higher region of the cortex, another from the septum. Lynch told Nauta that he planned to cut the main cortical input; this should cause a layer of hippocampal neurons that is connected to the pathway to degenerate. In this vacuum, Lynch hoped, neurons from an adjacent layer, which is connected to the input from the septum, would sprout new synapses, replacing the connections that had died.

The trick was detecting the new connections, which would be difficult, if not impossible, to distinguish from the surrounding tissue. After all, performing the experiment involved killing the rat and slicing up its hippocampus like so much Swiss cheese. There was no way to tell before from after—to know what the hippocampus looked like when the rat was still alive, before the experiment was done. But Lynch had an idea. Unlike the rest of the hippocampus, the neurons in the pathway from the septum were known to be cholinergic, that is, they use a chemical called acetylcholine as a neurotransmitter. The other neurons seemed to communicate using a chemical called glutamate. If sprouting occurred, the newly grown connections should contain acetylcholine, not glutamate.

Included in the neuroanatomist's palette is a dye that stains for an enzyme, acetylcholinesterase, that is always present in the cholinergic neurons; it is the chemical the neuron uses to break down excessive amounts of the neurotransmitter. Using the dye, Lynch believed he could make the newly sprouted neural connections show up as a dark band on a microscope slide.

Nauta was skeptical.

"He said to me, 'That's a fantastic experiment, and those kind of experiments of course never work.' And I said, 'I know, but I've got to try it.'"

Lynch's reputation as a scientist does not rest on his ability as a skilled experimenter. He hates the tedium, preferring to theorize while others do the dirty work; then he delights in poring over the results, looking for patterns. So Lynch designed the sprouting experiment and told one of his undergraduate students, a woman named Sarah Mosko, how to carry it out. First, she visited another lab to learn the staining technique. Then she lesioned the rat brains in the appropriate spot, allowed time for sprouting to occur, and killed the animals. After removing their hippocampuses, she froze them to make them solid, then sliced them with a type of precision saw called a microtome. Then she stained the slices, finally mounting them on slides.

Gary Lynch clearly remembers the evening in 1971 when Sarah Mosko came to him in the hallway with one of the slides. He held it to the light and saw a band that, if his interpretation was correct, represented newly grown neural connections.

"I absolutely couldn't believe it," Lynch said some fifteen years later. "You could see the effect with the naked eye. It was that big. And this was the very experiment that Nauta had told me would never work. The great man himself had told me that. And I could see the thing. And, as they say in the trade, as far as my career was concerned, the rest is history.

"You can imagine what it was like when I got seven reprint requests for my magnum opus on arousal. You know what it's like to get seven *hundred* reprint requests? You know what it's like to have the phone suddenly ringing off the hook, to be written up in the London *Times*, to have the television networks calling you?"

As important as the discovery itself was the feeling of having glimpsed through the veil of maya at nature's machinery humming underneath.

To confirm his interpretation, he decided to see if he could make the sprouting effect go away. He repeated the experiment with another rat. But after lesioning the input line from the cortex and allowing time for the connections to form, he cut the other input, the one from the septum. This should make the new synapses wither and die. Sure enough, when the brain was sliced and mounted, the telltale band was not there. Further experiments showed that what had grown were genuine, functional synapses. To this day, he still marvels at the effect.

"It's so hard to explain what happens to you at that moment, when you make a discovery like that. I don't want to sound pompous, but you never lose the feeling that you can do it again. And you also get a sense of what a real result is—reality. And once you have that, you're always after that same level of understanding.

"You don't need statistics, you just show a picture. Any anatomist could pick it up and look at it and say, 'Holy Christ, he built a new connection.' And that's exactly what the result said: *We build new brain circuits.*"

While his results were being heralded as new hope for the brain-damaged—a prediction that has turned out to be sadly premature—Lynch was becoming intrigued by another possibility. Studies of amnesiacs had shown that the hippocampus is vital to learning. Patients who have suffered hippocampal damage lose their ability to form memories. While they can recall what they learned before the damage, they can't acquire new facts. Perhaps sprouting was not simply a response to brain damage but a mechanism for growing the circuits in which memories are stored, for creating the patterns of neurons that serve as the symbols for what we know.

"So the thought began to grow: Maybe this sprouting of synapses, this capacity for making connections, does not only occur during injury or after injury—maybe this is a capacity we actually use," Lynch said. "Maybe all this plasticity is there as a learning mechanism! And that grew and grew and grew as an idea.

"People were stunned to find that their brains could grow new wires. But the irony, I think, is that in the end this is going to be infinitely more important, because it's saying not only can your brain grow new wires and connections, but you're doing it moment to moment."

Sprouting, he believed, or something like it, goes on all the time.

Strengthening Connections

OF COURSE, it was not enough to show that under certain artificial circumstances sprouting could be made to occur in the hippocampus. To show that the phenomenon had something to do with memory, Lynch needed to show that the brain could change in response to information, the electrochemical signals that are used to send messages from the senses. In 1973 some of the evidence Lynch was seeking appeared in a paper describing the work of a British neurophysiologist named Timothy Bliss and his colleague Terje Lømo. Using patterns of electrical pulses, they reproduced something that seemed very much like a memory trace, one of Karl Lashley's engrams.

Using an anesthetized rabbit, Bliss and Lømo stimulated a neural pathway leading to the hippocampus with a single electrical pulse. Then they measured the resulting voltage generated further down the line. At first, this output voltage was quite low, indicating that the synaptic connections in the circuit were very weak. But by repeatedly stimulating the pathway with high-frequency bursts of electricity, they were able to somehow turn up the volume of the connections. Now, whenever the input was stimulated, the neurons further downstream would vigorously respond. The small piece of brain tissue had learned a new trick. Weeks later, it retained the ability.

The scientists called the effect "long-term potentiation," or LTP. An event lasting several seconds had caused long-lasting neurological change.

"This phenomenon, as you might imagine, caused enormous excitement," Lynch said. "In some sense it was a realization of a dream that physiologists long have had. Here they saw something that looked like memory, and they produced it."

But Bliss and Lømo had simply shown that high-frequency electrical stimulation somehow made neural circuits more potent. They had no idea what the underlying mechanism was.

In a series of experiments, Lynch studied the effect in more detail. Recently, two neurobiologists, Chosaburo Yamamoto and Henry McIlwain, had invented a method for keeping slices of brain tissue alive in dishes. Using his own simplified version of the technique, Lynch's laboratory induced LTP in slices of hippocampus, allowing the effect to be studied with a precision that was impossible when working with a whole anesthetized animal.

Lynch assumed that LTP was causing synaptic changes. Either new connections were being formed, or there was some kind of strengthening of the ones that were already there. But other possibilities had to be ruled out. Perhaps the neurons themselves were changing, not just their synapses.

"Bliss didn't know whether the changes were selective to the synapses," Lynch said. "We designed an experiment in which we had several inputs colliding on the same cells and we potentiated one set of inputs and showed that the others weren't affected. So I convinced myself that the target cells themselves weren't changing somehow, that the changes had to be at the synapses that had been stimulated. It's one of the things that is axiomatic in the field now. It's a terrible thing to go to a meeting and hear people stand up and say, Oh, well, the changes are known to be selective to the synapses that have been stimulated. And you're going, Wait a minute, wait a minute; that wasn't given by Moses. That was a very difficult experiment."

But what was it about the synapses that had been altered? There were several ways that LTP might strengthen neural connections. It might be a presynaptic effect, one that causes the neurons sending the signals to release more neurotransmitter, as in Eric Kandel's snails. Or the effect might be postsynaptic: The amount of transmitter would remain the same, but the neurons on the receiving end would somehow become more sensitive, perhaps by increasing the number of receptors, the structures that respond to neurotransmitter. Or a combination of pre- and postsynaptic changes might be occurring. In addition, the brain might actually be growing entirely new synapses between the cells, creating new circuitry. Lynch found this last idea particularly attractive: "My immediate thought was that Bliss had triggered the sprouting mechanism."

An Exotic
Phenomenon

IDEALLY ONE WOULD BE ABLE to test Lynch's hypothesis about memory by inducing LTP in a neural circuit, then zeroing in on a single neuron to see if its synapses had changed, or if new ones had sprouted. Unfortunately, technology doesn't allow anything near this level of precision.

Instead Lynch had to use a brute-force technique. He spiked pieces of hippocampal tissue with electrodes and tried to stimulate as many neural pathways as possible. Then he and his students sliced the hippocampus into pieces thin enough to be photographed under an electron microscope. After making hundreds of these micrographs, as they are called, he taught several of his undergraduates how to read them, marking the synapses with felt-tip pens. Mixed in with the micrographs of tissue that had undergone LTP were an equal number of photos of normal slices. To guard against experimental bias, the students weren't allowed to know which were which, but the micrographs were coded so that after they had been marked they could be separated into two piles and compared.

After a year of this tedious work, Lynch was depressed to discover no difference whatsoever between potentiated and normal tissue. The control group and the experimental group looked exactly the same. LTP seemed to leave no mark after all; there were no engrams to be found. Then one night, just as he was getting ready to catch a plane to a conference, he was thumbing through stacks of micrographs yet another time hoping to see a pattern. Suddenly he noticed that, in addition to the synapses that had been marked according to his instructions, many of the micrographs had red marks in one corner. Sometimes there was one, sometimes two or three. And there were more of these strange markings on the micrographs in the LTP stack than in the normal one.

Lynch felt a stirring of anticipation. He tracked down the student, Michael Oliver, who had done the marking, and asked him, "What have you got these red marks up here for?" And he replied, "Well, you told me to look for this, this, this, and this, and we marked over six different things with that mark. But there's this other thing you didn't tell me to look for. And I just didn't know what to do with it. So I put a red mark up there whenever I saw one of them."

Lynch stared at Oliver with a mixture of excitement and disbelief. "I asked him, *'What are you marking? What do you see that causes you to put that red mark up there?'* And he said, 'Oh, let me show you. Right there. Those things.' And indeed it was a kind of synapse that I had never thought to tell him to look for."

In the mammalian brain, most synapses form on little bumps on the dendrite called spines. Lynch's student had unknowingly been marking something called shaft synapses, in which there is no spine.

"I frantically went through all the data, and sure enough, those kinds of synapses had increased by 35 percent," Lynch said. "The statistical significance was beyond imagining. I had to go catch a plane, and I'll never forget. I was thinking, God, this can't be true, this can't be true! We've finally found something.

"It was the first evidence that a physiological activity lasting for a fraction of a second is, within minutes, creating synapses. And at the time I couldn't believe it. And to this day I find it amazing. I thought it would be much more subtle.

"An adult neuron was sitting there with the capability of making new axon terminals, new receptors—the whole apparatus could actually be built from the ground up."

Many of Lynch's colleagues found it even harder to believe. The evidence was, after all, quite statistical. Again, there was no way to do a before-and-after experiment on the same piece of hippocampus, showing that it had more synapses after LTP. The only way to count synapses was to slice and mount tissue on a slide. By then it was already dead meat. No amount of stimulation would make synapses grow. To detect change one had to compare two groups of tissue— stimulated and unstimulated—and see which, on average, had more synaptic connections. This involved examining many, many slides. Looking at any single pair of micrographs, one could detect no obvious change.

"I could show you an electron micrograph, and you wouldn't know the difference between the control and the experimental tissue," Lynch said. The experiment simply indicated that brain slices that had undergone LTP were, in general, more likely to contain shaft synapses. There was no way to know whether the synapses had actually appeared because of LTP, or that they were even in pathways that had been electrically stimulated. "The assumption is that the ones we were looking at were connected to the axons that were stimulated," Lynch said, "but that's an assumption, a leap of faith."

Few of Lynch's colleagues were ready to take what seemed like a Kierkegaardian leap into the absurd. After all, it was generally assumed that to produce the kinds of structural changes that encode memory, new proteins would have to be synthesized. Somehow a message would be sent to the genes in the neuron's nucleus, located in the body of the cell, telling the DNA to produce the proteins and enzymes necessary for building membranes, receptors, all the apparatus required to make synapses. The reaction Lynch believed he had observed was too swift to allow for all this activity.

"There was no time for protein synthesis, no time for the genome to be activated," he said. "Our effect was present in ten minutes. And we didn't look at thirty seconds. For all I knew the synapses were there at thirty seconds. Well, the reviewers' reaction to this was, uniformly, 'No way.' " The findings were initially published as an obscure three-page note that went largely unread. Later a more sizeable paper appeared in the *Journal of Neurophysiology*. But people were less interested in the experiment than in the technique Lynch had used to keep the brain slices alive.

"Everybody wanted to know about the brain slices!" he said. "LTP hadn't really caught fire, and among biologists memory was still disreputable. It's very hard for people to believe this now, how recently it's been that the biology of memory wasn't spoken about in polite company. You didn't bring the subject up. And, God knows, you would never mention the relationship between an animal form of memory and human memory. That was even more *verboten*.

"The biology of learning and memory in mammals was almost a nontopic in 1979. You have to remember there was scotophobin. You remember scotophobin. And there was the Planaria, and there was the RNA transfer.

"Kandel, more than any other individual, was making memory a proper field of study for a neuroscientist. And that may have been another factor. That was at the height of the invertebrate rage. Everybody was doing invertebrates. The world was just afloat with Aplysia and crabs and lobsters and insects—everybody had an insect model of learning. See, that's what the *real* scientists were doing. The mammal people were in total disarray and scattered across the planet. I don't think I'm exaggerating too greatly. It was an amazing time. So I don't think my work had the impact it might have had, and that was an extremely disappointing thing for me. In fact I don't think to this day it's had the impact it should have."

If Lynch was to be believed—indeed, if he was fully to convince himself—he would have to find a chemical explanation of what was happening. What kind of molecular reactions could be causing the synapses to form? There was little in the repository of biochemical knowledge for him to rely on.

"Memory, after all, is a very exotic biological phenomenon," he explained. "The biological world is a world of homeostasis. It's a world in which if you produce one change, there'll be something there to correct it. So if you eat a meal and your body temperature rises, there is something to correct that, to bring it back down. If you take it on down to the cellular level, if you add a phosphate group to a protein, there will be an enzyme there to dephosphorylate that protein. So now you come along and say, Well, now I have an experience that creates a pattern of activity in my brain that produces a change that lasts for years or decades. That's exotic biologically. What kind of chemistry could produce it? Those were mysterious questions."

The Porcupine Effect

SINCE THERE WAS NO WAY to gaze into the cellular machinery and watch the synaptic chemistry unfold, Lynch had to approach the problem indirectly. He reasoned that if long-term potentiation—and, by implication, learning—were really causing the brain to grow new synapses, it would have to produce more receptors, the proteins that respond to the neurotransmitter. Increasing the number of receptors would also provide a means for turning up the volume of existing synapses. The problem was figuring out how to count them.

In 1978 Michel Baudry, a tall, amiable French neurochemist, joined the laboratory at Irvine. Together, he and Lynch worked out a strategy for showing that LTP led to the appearance of new receptor molecules. After using electrodes to electrically stimulate a hippocampal slice, they would liquefy it in a blender. Then they would use a centrifuge to separate out what are called synaptosomes, the parts of the dendrites and axons that clamp together to form synapses. What they would do, in effect, was to take the complex chemistry of the synapse and distill it from the rest of the neuron into a test tube. This method allowed them to perform experiments that were impossible in an intact brain.

The next step was to add a chemical called glutamate to the broth.

Glutamate is the amino acid that is believed to be the main neuro-transmitter used by the hippocampus to send signals from cell to cell. In the broth, glutamate should latch onto the glutamate receptors, just as it would in an intact synapse. By measuring the amount of this chemical binding, they could tell whether the number of glutamate receptors in a piece of brain tissue had actually increased. Beginning in the late 1970s, Lynch and Baudry showed that LTP indeed seemed to produce more glutamate binding. Stimulated tissue seemed to have more receptors than unstimulated tissue.

But what was the electrical stimulation actually doing that could cause new receptors to appear? Answering this question called for some more complex laboratory maneuvering.

Lynch and another colleague, Tom Dunwiddie, had learned from earlier brain-slice experiments that without enough calcium in the fluids bathing the neurons, LTP would not occur. In fact they found that they could block the LTP effect by injecting brain tissue with a calcium chelator, a chemical that tied up calcium. And so, Lynch was led to suspect that the first step in the LTP chain reaction occurred when an influx of calcium poured into the neuron from the surrounding intercellular fluid. Then the calcium would somehow trigger the synaptic change that led to the laying down of engrams.

To test this idea Lynch needed to cut out the first link in the causal chain and see whether calcium alone—with no previous electrical stimulation—could make new glutamate receptors appear. By putting calcium into a broth of synaptosomes, he found that he could indeed increase what seemed to be glutamate binding, just as though he had administered LTP. It was seeming more and more that calcium was the catalyst for the formation of memories. Lynch became even more convinced when he discovered that the effect was irreversible. Remove the calcium from the mixture after the reaction had occurred, and the binding effect remained. Another piece of the puzzle was in place—calcium set off the reaction, but it wasn't necessary to sustain it.

Now he just had to figure out what the calcium was doing. How could it cause rapid, irreversible change?

"We wanted a chemical mechanism that lived in the synapses, that could produce an effect in a matter of seconds that would last indefinitely," he explained. "And indeed there is one class of enzyme found throughout the body that produces irreversible effects. These are something called proteases. You can do a lot of chemistry—you can phosphorylate things, you can methylate things, aceylate things. All those reactions are fully reversible. But proteases break proteins, and that's like Humpty-Dumpty—once broken they can't be put back together again."

As he read through the literature, Lynch found what he had hoped was true: There was a type of protease, called calpain, that was activated by calcium. It had been known since 1964 that brain tissue from rats contained calpain. At the time, this destructive enzyme was being studied because it seemed to be involved in the degeneration of muscle and nerve tissue. But no one had ever suggested that a molecule whose job was to break proteins could somehow cause the kind of changes that led to the formation of memories.

As he pondered where else in the body sudden structural change occurs, Lynch came up with an unlikely parallel to memory formation: the clotting of blood. As part of this complex process, blood cells called platelets change their shape from disklike and smooth to rough and spiny, allowing them to hook together and help seal a wound. This sudden change occurs when the calcium level in the cell is increased. A similar reaction also occurs in erythrocytes, or red blood cells, and might be involved in their ability to change shape and squeeze through tiny capillary openings.

"I can show you pictures of blood cells, what happens to them when that little action occurs, and it's horrifying to see it go from something that looks like a Frisbee to something that looks like a porcupine in about thirty seconds," Lynch said. "It is quite a remarkable kind of thing."

But how could a cell change so quickly? Lynch believed he had an explanation. Contrary to many people's mental imagery, the membrane that forms the wall around a cell is not a solid structure. It has the viscosity of thirty- or forty-weight motor oil. Its rigidity comes from a chemical scaffolding called the cytoskeleton, a protein frame that,, as Lynch describes it, the membrane is dripped onto. Lynch believed that clotting occurs when calpain eats through the platelet's cytoskeleton, allowing the cell to change shape.

A platelet is just about the size of a synaptic spine. As Lynch examined the before and after pictures of blood clotting, he imagined the same kind of thing happening in the brain. When a neuron is stimulated by the high-frequency impulses of LTP, he speculated, microscopic channels open in the membrane, allowing calcium to flood into the cell. Calcium activates calpain, which eats the proteins that form the cytoskeleton. This would allow glutamate receptors that were hidden deeper in the membrane to pop out. Thus the synapse would become more sensitive to glutamate. In other words, its volume would be turned up. The calpain machinery seemed to be in the platelet. Perhaps evolution had adopted this fundamental device when it needed a way to produce sudden change in brain cells.

Breaking the cytoskeleton could occur almost instantaneously, accounting for the rapid LTP effect. With repeated stimulation, more radical change might occur. Patches of newly exposed glutamate receptors might migrate and cluster in new locations on the dendrite, forming the receiving ends for entirely new synapses, new circuitry. Perhaps strengthening existing synapses accounted for short-term memory, which was later stored for the long term by the sprouting of synapses.

Learning, like the clotting of blood, might be the result of the porcupine effect, these infinitesimally tiny explosions. As we listen to a lecture or read a book, little bumps on our dendrites would swell and explode. The result would be new circuitry. Whether by increasing the level on the synaptic volume controls or by forming entirely new synapses, neurons would make links with neurons that they had not communicated with before.

In experiment after experiment, Lynch and Baudry compiled evidence for the hypothesis. "What we discovered was that calpain indeed breaks the cytoskeleton," Lynch said. "Calpain takes the very molecules that form it and breaks them in half."

Some experiments were carried out with brain slices, others with the broths of liquefied synapses. It was like playing a game of twenty questions, interrogating an unknown phenomenon. Each experiment, which could take days or weeks to complete, would answer a single question.

Q. Is calpain truly essential to the reaction?

A. When leupeptin, a protein-like substance known to block the action of calpain, is added to the broth, it seems to discourage increased glutamate binding. If Lynch's interpretation was correct, this would mean that with calpain turned off, new receptors don't appear.

Q. Granted that calpain and calcium seem to be necessary ingredients to the reaction. But can we be sure which triggers which?

A. If leupeptin is added to the broth before calcium, then glutamate binding is interrupted. If added after the calcium, it has no effect—apparently the calpain has already done its job, initiating synaptic change.

Enlisting an ever-changing team of undergraduates, graduate students, and postdoctoral researchers, Lynch and Baudry laid the foundation for the theory, brick by brick. To show that calpain was actually breaking the proteins that made up the cytoskeleton, they used a technique similar to what is called chromatography. The name comes from the fact that if droplets of certain inks or dyes are put on blotter paper and hung up to dry, they will spread and separate into their component colors; the heavier chemicals will be pulled farther down the paper than the lighter ones, leaving a spectrum of colored bands. To separate proteins, scientists use a gel instead of blotter paper, and instead of relying on gravity they use an electrical current to propel the molecules. By measuring the location of the bands, they can roughly identify molecules of different sizes.

Again using a centrifuge, Lynch and Baudry produced a broth rich in synaptosomes. When they broke apart the solution with the chromatographic technique—called gel electrophoresis—a band appeared at about the spot where one would expect the skeleton molecule, called spectrin, to come to rest. But if calpain was added first to the broth, the band would be less intense. More exciting still, another band would appear that seemed to represent fragments of the broken spectrin. Again, leupeptin, the calpain inhibitor, blocked the effect.

As a body of indirect evidence for the calpain hypothesis began to accumulate, Lynch considered its weak points, the links in the logical chain that would be most vulnerable to attack. While it seemed clear that calpain was causing the breakdown of a protein that lived inside synapses, gel electrophoresis was simply not precise enough to show beyond doubt that it was spectrin. Many proteins have similar molecular weights. Perhaps calpain was causing the snapping of proteins that had nothing to do with structural change.

One way to identify a protein is to mimic the body's immune system, which has an uncanny ability to search out and destroy very specific invaders. It does this by identifying molecules on the surfaces of viruses and other interlopers and manufacturing antibodies—complementary molecules shaped to latch onto the invaders' molecules, much as neurotransmitters latch onto receptor molecules in a synapse. In the case of the immune response, this binding begins a series of biochemical events that destroys the invader.

By injecting spectrin into rats, Lynch's colleague Robert Siman used the rodent's immune system as a chemical factory, producing antibodies to the molecule. Then he labeled the antibodies with radioactive isotopes and added them to a broth containing spectrin. The antibody molecules would seek out the target, latching tightly to it. Now the spectrin itself was radioactively tagged, making it much easier to identify. When he added calpain to the mixture, he was able to show that it had a strong affinity for spectrin—further evidence that when calpain was unleashed in a brain cell by LTP, its role was to break cytoskeletons, initiating changes in the brain.

And so it went. Experiment by experiment they seemed to be zeroing in on their target. Still, there was no way to watch what a synapse did when it was jolted by LTP, to actually see calcium flowing into the dendrite, activating calpain, breaking spectrin so that glutamate receptors popped out. But like a lawyer preparing to defend a dubious client, Lynch continued to marshal his evidence, weaving it into a convincing tale. The evidence was all circumstantial, and there were a number of ways to explain it. But part of Gary Lynch's talent as a scientist is connecting the bits and pieces of seemingly unrelated data into a good story. Not that there is anything disingenuous about that. It is part of the essence of theory-building.

Sharing a Hallucination

AT THIS POINT it seemed plausible that Lynch was looking at something that really existed. But it was hard to tell. In the complex world of brain chemistry, all would be chaos unless researchers ignored most of what is going on, dismissing it as background noise, and focused on the one or two reactions that they have chosen to study. When scientists view the brain at the level of chemistry, they are quickly overwhelmed by detail. Serotonin, dopamine, norepinephrine, histamine, gamma-aminobutyric acid—there are more than two dozen different substances that act as neurotransmitters. When a neuron sends a message, a medley of sodium, potassium, and calcium currents flow through channels in the membrane of the cell. This complex chemical dance can't easily be teased apart without doing it violence. In the spirit of Heisenberg and his uncertainty principle, the observation disrupts the experiment.

How do you separate foreground from background, the essential from the superfluous? One can imagine alien anthropologists studying an earthling computer, trying to determine how it thinks. They might spend years analyzing the plastic case surrounding a microchip, noting in careful experiments that during the machine's operation, molecules of plastic twist from the tiny currents of heat; further experiments would show to what degree these thermal fluctuations were generated by the chip itself or by the transformer on a nearby power supply. They might theorize that the plastic is doing the information processing, that information is stored somehow in the configuration of the polymers, or that the cooling fan is an integral part of the machine's mental life. Lynch had a hunch he was on to something, but for all he knew he was devoting years of his life to an interesting biochemical effect that had nothing whatsoever to do with memory.

It has only been in the last few decades that scientists have worked out their elaborate theory of how neurons communicate. While it often suffices to gloss over the details and simply remember that axons send signals by squirting neurotransmitter onto dendrites, it is impossible to fully appreciate the difficulty of this business without considering how overwhelmingly complicated neuronal chemistry really is. After thousands of experiments, the community of neuroscientists have agreed on a story that goes something like this:

The first step in the chain reaction occurs when a neuron fires in response to some kind of input. Most of the time the stimulation comes from another neuron sending out neurotransmitter. But in the case of sensory neurons, the messages can come directly from the environment: Neurons in the skin respond to pressure, neurons in the retina respond to light. In any case, once the neuron fires, an

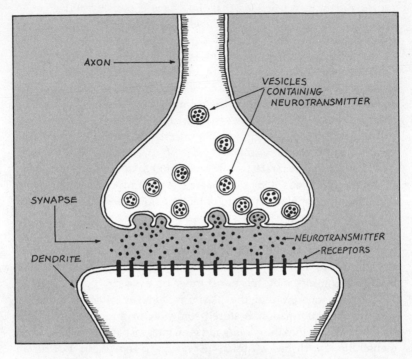

CLOSEUP OF A SYNAPSE.
The axon of the sending neuron signals the dendrite of the receiving neuron by releasing neurotransmitter into the synaptic gap.

electrical signal travels the length of its axon until it reaches the end—a cul-de-sac called the axon terminal—and bursts little bubbles, called vesicles, which are filled with molecules of neurotransmitter. From here the neurotransmitter is released into the synapse, where it binds to the receptors embedded in the dendrite of the neuron on the other side of the gap. This binding action stimulates the receiving neuron by causing channels to open up in its membrane, allowing positively charged sodium or calcium ions to enter the cell. An ion is simply an atom in solution that carries an electrical charge.

At the same time, other neurons are also feeding signals to the dendrite through other synapses, opening still more ion channels. Bit by bit a positive charge builds on the dendrite as more and more ions flow in. If enough of a charge accumulates, the neuron fires, sending its own spike hurtling toward the next set of vesicular balloons. This summation of signals occurs in both space and time: Enough signals impinging at once from many other neurons might fire the cell, or many signals from a single neuron might arrive so rapidly that the charge becomes great enough to trigger the firing of the action potential.

But this is only part of the story. Not all of the signals a dendrite receives are excitatory. As the positive charge is building, another kind of synapse reacts to neurotransmitter sent by other neurons by letting negatively charged chloride ions into the cell. These inhibitory signals cancel out the positive, excitatory signals. So the neuron is constantly teetering on the brink of firing, pushed one way and then the other by competing signals.

.   .   .

Viewed at that fairly abstract level, the process seems intricate enough. But delve a little deeper, and the complications mount. How is it that a preponderance of positive charges can cause a brain cell to fire? Again, this can be explained with chemistry.

Like any cell, a neuron is surrounded by a membrane. This fatty wall is designed to strictly control the kinds of molecules and ions that flow in and out. In the case of the neuron, the membrane contains not only ion channels, molecular gates that open and close, but also submicroscopic pumps. These molecules constantly transport sodium ions to the outside of the cell and potassium ions to the inside. While the subtleties of the process are dauntingly complex, the important thing to remember is that the result of all the pumping is to keep the cell slightly more positive along the outside of its membrane and slightly more negative along the inside. In other words, an electrical pressure, or voltage, is built up along the cellular wall. It's a very delicate balance. Molecules tend to move from areas of high concentration to areas of low concentration; and, of course, positive charges are attracted by negative ones. Given a chance, the sodium ions would love to rush back in.

This is exactly what happens when the neuron is pushed into

firing a spike. Sodium ions that had been crowding on the outside flood into the cell, carrying a spike of positive electricity down the length of the axon.

The key to this effect is another kind of channel that responds to changes in membrane voltage. While the receptors at the synapses are chemically gated—their ion channels open when they are activated by neurotransmitter—these other channels are said to be electrically gated. They open not when they are stimulated by another molecule but when the membrane that surrounds them reaches a certain voltage.

The length of the axon is studded with electrically gated channels. If enough positive charge builds at the dendrite, the change in voltage trips the first channels in line. They open, allowing sodium ions to flood into the cell. This spreads the positive charge slightly farther down the axon, opening more channels, which let in more sodium ions and spread the charge farther still. This rolling loop of positive feedback can be thought of as a tiny flame moving down a fuse; it propagates by heating the pathway in front of it, raising, point by point, the combustion temperature.

This voltage spike is not like a current of electrons traveling down a copper wire. It is a complexly orchestrated chain of chemical reactions. The spike is carried not only by the domino effect of the sodium channels; intertwined with this process are the actions of other channels, which expel potassium as the sodium comes rushing in. When the spike finally reaches the axon terminal, still another kind of channel opens, letting in calcium ions. Calcium is an integral ingredient in another set of reactions that leads to the bursting of the vesicular balloons. They fuse with the membrane of the axon tip and release their neurotransmitter into the synapse leading to the next neuron, where an identical process is waiting to be ignited.

Once it has fired a wave of spikes, a neuron is spent. Then the ion pumps go to work. Sodium is pumped back out of the neuron; potassium is pumped back in. The clock spring of the mechanism is wound up again; the disequilibrium is restored, awaiting another chance to be upset.

· · ·

It is sobering to realize that despite the impression given by drawings in science books (including this one), no one has actually seen all of this happen. The picture has been built up, inference by inference, over the last hundred years. Using a voltmeter or an oscilloscope, we can capture a neural spike as it is occurring, but we can't zoom in close enough to watch neurotransmitter molecules activating synaptic receptors, or ion channels clacking open and closed like calliope pipes as they carry the signal down the axon. To examine such tiny structures, we must slice and mount thin sections of dead tissue and put them under an electron microscope. At such high magnification, there are indeed vesicles that look remarkably like little balloons. But receptors, pumps, and ion channels are best thought of as metaphors; they are not so much devices as molecular processes too tiny to be seen.

Each of these structures is, in fact, a single protein molecule, a long twisting chain of amino acids designed to carry out certain very specific tasks. Like all proteins, a receptor has binding sites, patterns of charges that match complementary patterns on other molecules. This lock-and-key configuration is what allows a neurotransmitter to latch onto a receptor. When the two molecules join, the receptor is deformed; it flexes in just such a way that it is more likely to allow ions into the cell. The electrically gated sodium and potassium channels are proteins that twist and open when they are transformed by a voltage change.

None of this is as mechanical as it sounds. There is no little hinged cap covering a tunnel into the cell. Actually, the molecules that act as channels and pumps are constantly oscillating between various configurations. When a receptor binds with a transmitter, or a sodium channel is under the influence of a positive charge, the molecule is more likely to keep the shape that is most porous to ions. A neuron is mechanistic in the sense that a thunderstorm is. It is constrained by natural laws, but a great deal of chance is involved.

Forget for now the details that must be glossed over in any description of a biological process: that the cell is constantly manufacturing and replacing proteins—ion channels, pumps, receptors—and miraculously transporting them to the proper sites; that it is constantly breaking sugar into energy to power these reactions. The French neurobiologist Jean-Pierre Changeux has spent a good part of his life

studying how a single molecule, the receptor for the transmitter acetylcholine, operates. Even viewed in the most simplified schematic
form, neuronal firing is an overwhelmingly complicated process. If
only it were a fraction as simple as has just been described.

All theories are, in a sense, cartoons. There is no reason to suppose that people have evolved languages—either verbal or mathematical—that are fine enough to describe molecular events. Does it
really make sense to talk about a "bursting" of vesicular balloons or
a "spray" of neurotransmitter in an area less than a micron across?
There is always the danger of confusing the map with the territory.
It is never clear how close is the relationship between the action of a
molecule and the hard-won picture of it we hold in our heads.

Consider the case of the ion channel, which began simply as an
idea postulated in the early 1950s by the Nobel laureates Alan Hodgkin
and Andrew Huxley to explain how something like a neuron could
work. Wouldn't it be convenient, they suggested, if there were submicroscopic holes in the membrane that open and close only under
the proper conditions?

Over the decades, indirect evidence has accumulated to support
the belief in ion channels, but it has only been in the last few years
that scientists have developed techniques to study them. What seem
to be channel proteins can be isolated and inserted into an artificial
membrane where, lo and behold, they behave much as theory predicts. With an instrument called a patch clamp, it is now possible to
measure the activity of a single ion channel. A hollow glass electrode
with a tiny inner diameter is touched to a neuronal membrane, where
it forms a suction seal just large enough to cover one channel. Scientists still can't see the channel, but when they hook the electrode
to an oscilloscope, they can observe its opening and closing by measuring the ionic currents passing through. Again, channels generally
behave as theory predicts. When they don't, the theory is slightly
modified.

More clues about channels come from cloning and other techniques, which help scientists unravel the structure of molecules and
even build models that suggest how their shape can account for their
function. Still, all this evidence is circumstantial, and none proves that
ion channels exist. In this most ambiguous of sciences, it is never
clear what standard of evidence we should require, or even what we

mean by a fact. According to the classic view of science, data are gathered and analyzed for the patterns we call theories. But, to paraphrase Einstein, it is the theory that determines what facts we can observe. No one would have induced ion channels from the welter of data had not the theory of neuronal transmission put forward by Hodgkin and Huxley required that something like them exist. In the molecular universe of a single cell—where thousands of varieties of proteins are constantly being created and destroyed—there would have been no reason to seek out this one, seemingly obscure variety of molecular behavior, to declare it meaningful and separate it from the surrounding noise. But if a theory predicts a phenomenon that is indirectly confirmed by a number of different techniques, then we can tentatively agree to believe it exists. Bit by bit, this kind of evidence is accumulated and incorporated into our evolving mental pictures. Science, to quote the molecular biologist François Jacob, is a "constant dialogue between imagination and experiment."

Like much of science, the theory of the neuron is a shared hallucination, a network of self-reinforcing beliefs. By accepting a theory, scientists can move forward, gathering new evidence to test their assumptions. And so the theory continues to unfold. But it is always unsettling to think what we might be missing, for want of the tools and ideas that would let us see.

"One of the most difficult things about neurobiology is learning to live with ambiguity," Lynch said one day in his office.

It is almost as though the scientists living in the macroscopic domain of the laboratory were trying to decipher cryptic messages from another realm. Their patch clamps and oscilloscopes and electron microscopes are tools of communication, devices for capturing messages from a barely visible, alien world.

Artificial
Amnesia

To CUT THROUGH the thicket of biochemical ambiguities and show that he was on the right track, Lynch would have to demonstrate that disrupting the calpain mechanism not only interfered with LTP but with the actual storing of memories.

It was becoming more believable all the time that high-frequency electrical stimulation somehow strengthened (and perhaps created) synapses. And the assumption was widespread that the neural structures that resulted from this process were the long-sought engrams. Lynch and Baudry's test tube experiments seemed to suggest that calpain played a role in LTP. But maybe LTP was just a laboratory curiosity. The most dramatic way to show that Lynch was really unearthing a memory mechanism would be to see if blocking the calpain mechanism caused amnesia.

In 1983, Lynch and a Swiss scientist named Ursula Staubli surgically equipped rats with small pumps—the mechanical, not the molecular kind—infusing their brains with a constant supply of leupeptin, the calpain blocker. Then they tested the animals' ability to solve a problem in rat psychology known as the eight-arm radial maze.

Dropped into the middle of one of these simple worlds, a rat is faced with the choice of eight pathways that radiate from the central chamber like petals on a flower; at the end of each is a bit of food. Normal rats quickly learn to run to the end of a pathway, get the food, and move on to the next alley, without returning to chambers where the reward has already been eaten.

The rats given leupeptin seemed normal in almost every regard—there was no sign that they were sick or sluggish, that leupeptin was poisoning them. When put into the maze, they clearly recognized it as a familiar environment: They recalled the general shape of the problem, they knew there was supposed to be food at the end of tunnels. Knowledge already established had remained intact. But once they began to run the maze, the rats had a hard time remembering which paths they had visited, especially if they were removed in the midst of the experiment and returned several minutes or hours later. The leupeptin seemed to be directly interfering with their ability to form memories.

To see how universal this phenomenon was, Staubli tried another experiment, testing rats on how well they could learn to avoid an electrical shock. But this time, leupeptin seemed to have no effect. Why was a calpain blocker inducing amnesia in one experiment but not in the other? Looking for a way around this seeming contradiction, Lynch turned to his old college major, psychology. Psychologists had come to distinguish between two kinds of memory: procedural and declarative, learning *how* versus learning *what*. After we learn to ride a bicycle, we cannot articulate the knowledge we have gained; all the little micromovements are stored implicitly throughout our central nervous system as processes not facts. Some of this knowledge, if we can call it that, seems to reside locally in the form of spinal cord reflexes that are not under the direct control of the brain. In experiments with human amnesiacs and monkeys, Larry Squire, Stuart Zola-Morgan, and other researchers had built a convincing case that the hippocampus is crucial to the ability to store facts but not procedures. An amnesiac with a damaged hippocampus can learn simple skills, like reading print reversed in a mirror, but he cannot remember a thing about the training session.

As Lynch saw it, Staubli's experiments suggested not only that the calpain mechanism was involved in fact learning but that there were separate chemistries underlying the two forms of memory. The evidence was hardly conclusive. Much too little was known about leupeptin; it could be blocking learning by interfering with other cellular processes that had nothing to do with calpain. Still, the episode of the eight-arm maze provided a nice denouement to the calpain story. In 1984, confident that their ideas were solid enough to be revealed to the world, Lynch and Baudry wrote a paper for the journal *Science*: "The Biochemistry of Memory: A New and Specific Hypothesis."

A Hostile
Replication

THE ANNOUNCEMENT of Lynch's calpain hypothesis played to mixed reviews. "Skeleton Key to Memory?" asked a headline of an editorial written for *Nature*, the British science journal. "If, indeed, something as simple as the cleavage of a single protein represents a major and universal event in the process of learning, this would be cause for new hope, if not yet for dancing in the quad."

Most of Lynch's colleagues were more skeptical. They thought he was reading too much into the data, seeing a grand architecture that wasn't really there.

Over the years, bits and pieces of proposed memory mechanisms had popped up in the literature now and again. Their inventors speculated on the role of various enzymes in causing the synaptic changes associated with memory. Chemicals with forbidding names like calmodulin, protein kinase C, and cyclic AMP had become familiar presences to those whose careers were spent thinking about how learning might cause changes in the brain. But what in God's name was *calpain*? Except for researchers studying possible causes of muscle and nerve degeneration, no one else in neuroscience was thinking much about calcium-activated proteases. The calpain hypothesis was just too weird. When he talked about it, Lynch recalled, "People looked at me as though I had come down from the planet Xenon."

He also had an image problem to overcome. Despite Lynch's rather impressive list of publications, many neurobiologists found him a little hard to take. It was bad enough that his training had been in psychology. Even worse, his brashness and flamboyant behavior at neuroscience conventions had given him a reputation as something of a madman. At the annual meetings of the Society for Neuroscience, he was less likely to be found raptly perusing the aisles of posters describing the latest developments in LTP research than standing at the convention center bar. Clustered around him would be a gaggle of admiring colleagues and former students trading jokes and stories. Gary was a show in himself. In the evenings he would hold court in the lounge of his hotel, drinking late into the night as he expounded on his latest theories—not only on memory and neuroscience but on evolution, anthropology, history, literature, the way the world works. "Gary has a theory for everything," Michel Baudry observed.

While Lynch's stuffier colleagues avoided these sessions, others considered them the highlights of the conference. Fueled with ethanol, Lynch's quick mind would shift into overdrive, connecting facts and opinions into vast, sprawling theories that would shift and dissolve as quickly as they were formed. Lynch had launched his career with his work at Princeton on how the brain regulates its state of arousal. Sometimes it seemed that Lynch's own nervous system had lost this mechanism entirely; as in his electrically lesioned rats, the feedback loop that was supposed to keep his cerebrum in check seemed to be severed. His brain exploded with ideas, and there was little use in trying to steer the conversation.

To his detractors, science just seemed too easy and fun for Gary. He couldn't be making serious discoveries. And yet, he and his laboratory were becoming a force that was hard to ignore. As his strongest critics derided the calpain hypothesis, they tended to forget that it was Lynch and his colleagues who first showed that calcium was necessary to LTP. It was Lynch who first found evidence that LTP caused new synapses to sprout in the brain.

Some of the skepticism about this, the most radical of Lynch's claims, had begun to abate somewhat in the months before the calpain hypothesis was unveiled. At the 1983 Society for Neuroscience meeting, William T. Greenough, a psychologist at the University of Illinois who is as reticent and precisely spoken as Lynch is flamboyant, announced that he had confirmed the experiment indicating that LTP causes new synapses to appear.

Since the early 1970s, Greenough had been compiling a persuasive body of evidence that learning leads to new synaptic connections. As early as 1973, he had helped show that rats raised in stimulating, complex environments, in which they are allowed to explore, learn, and interact with their fellow rats, tended to have more dendrite per neuron than rats who lead duller lives. This, of course, would allow more area for new synapses.

But Greenough doubted that synapse formation could be as simple and straightforward a matter as Lynch wanted to believe. Working with a colleague named Fen-Lei Chang, Greenough performed a more refined version of Lynch's electron micrograph experiment, eliminating some ambiguities that had cast doubt on the results. In what he later conceded was a "hostile replication," Greenough was surprised to find that high-frequency electrical stimulation not only caused synapses to form, but that the effect happened in as quickly as ten minutes.

"I had been an advocate up until that time of the notion that there was a process of constant turnover of possible, or potential, synapses in the brain, and that what memory did was to sort of stamp them into place," Greenough later said. "These data aren't necessarily incompatible with that, but they at least make you think about the alternative: that synapses can actually form in response to the kinds of events that initiate memories."

One of the troubling things about Lynch's experiment was that while it found an increase in the so-called shaft synapses, which are relatively sparse in the hippocampus and other higher regions of the brain, there was no obvious increase in the more common spine synapses, the ones with little bumps on the dendrite. While they didn't see any new spine synapses, Greenough and Chang found an increase in a third kind, the so-called sessile synapse. These synapses had stubby dendritic spines that were smaller than those of actual spine synapses. Perhaps the sessile synapse represented a transition stage between the flat, featureless shaft synapses and fully developed spine synapses. The implication, then, was that a little learning causes shaft synapses to form. As the memory is solidified, the shaft synapses might slowly begin to develop into ones with spines.

Thanks to Greenough, Lynch was no longer a lone voice. Another laboratory had found that LTP seemed to form new connections at an astonishing speed. But, Greenough said, "that doesn't mean we buy the rest of the package." He said he was "decidedly neutral" about the calpain mechanism. "It's one of several hypotheses," he said, carefully choosing his words. "It certainly has some evidence in support of it. There are certainly many additional aspects of it that would have to be demonstrated before we could really accept it, and there are other hypotheses as well that deserve consideration."

The King
of the Hill

OUTSIDE Lynch's Orange County domain, this guarded kind of reaction seemed to be the best he could hope for. During the next few
years he found his theory of memory besieged on many fronts and,
on a few occasions, in danger of lapsing into disrepute. It didn't help
matters that the mechanism he was proposing bore no resemblance
whatsoever to one being put forward by Eric Kandel.

Kandel was widely regarded as one of the most powerful people
in neuroscience. In the years since his theories of memory in the sea
snail Aplysia had become firmly mortared into the foundation of neurophysiology, he had joined the small pantheon of scientists whom
everyone in the field at least grudgingly respected: David Hubel,
Torsten Wiesel, Vernon Mountcastle, Walle Nauta, John Eccles. The
prestige he had gained from his demonstrations that changes in behavior were mirrored by changes in neurons had attracted an ever-
growing influx of disciples and grants. He now sat at the top of one
of the most lavishly financed neuroscience research institutes in the
world.

As the head of the Howard Hughes Medical Institute's Center for
Neurobiology and Behavior at Columbia University, he directed a staff
of researchers, many of whom were staking their careers and repu-
tations on refining and expanding the general theory launched by
Kandel: Learning causes a change in the amount of neurotransmitter
a neuron releases. Gary Lynch was claiming just the opposite: that
neurotransmitter flow probably remains steady while the number of
receptors increases. Each was convinced that the other was utterly
wrong.

If the biochemical reactions Kandel had found in his Aplysia were
indeed what he liked to call "letters in the cellular alphabet of
learning," his followers were hoping to weave them into words, sentences, a whole language that would tell the story of how learning

causes changes in the brain. Some of Kandel's colleagues talked about his laboratory on Manhattan's far Upper West Side as though it were a factory: In room after room, workers sat at benches churning out results that would bolster Kandel's version of how memory works. When they left Columbia to colonize other laboratories, they took with them the ideas of Eric Kandel.

. . .

Born in Vienna, Kandel grew up in the Flatbush section of Brooklyn. Four years at Harvard University, where he studied history and literature, helped give him the poise—some call it arrogance—that is supposed to come with an Ivy League education. Kandel is inevitably described as articulate and refined. But this cool self-control is precariously balanced. The folklore of neuroscience is filled with tales of shouting matches between Kandel and some benighted soul who dared to question one of his theories. This volatility could come as a shock. One minute Kandel might be talking about his high-minded view of science as a cooperative endeavor—he hated hearing other scientists referred to as "competitors." The next minute he would be deriding one of the brethren as a "nut," another as the "Fawn Hall" of neurobiology. It was startling, the quickness with which he could shift from being charming to being insulting and rude.

Stories about Kandel's excesses must be eyed with a certain amount of skepticism. Power always generates resentment and even paranoia. In addition to the Gary Lynch drinkathons, another common occurrence at neuroscience conferences was small knots of younger researchers complaining, sometimes bitterly, about Kandel. The griping would generally be sotto voce; it was assumed that if Kandel didn't like you, he could ruin your career. Why, you never knew when he or, more likely, one of the many scientists who owed him fealty would be asked to review your paper for a journal or evaluate your grant proposal. What if one of his minions was sitting at the next table, listening in?

It was fear as well as respect that caused even his enemies to preface their complaints with grudging praise: "Eric is one of my heroes, *but* . . ." or "Eric is truly one of the great men of neuroscience, *however* . . ." Then the complainant might go on to express incredulity at the way Kandel has parlayed an admittedly impressive body of work about adaptive behavior in sea snails into a scientific kingdom. No one doubted that he was a careful, talented researcher. Yet some of his colleagues attributed his success less to the merits of his theories than to his abilities at intellectual salesmanship. Why has science been so ready to accept that a set of biochemical events occurring in an obscure creature hardly more sophisticated than a plant can tell us something important about the way the human brain works?

"Eric is far and away a very classy intellect," said one neurobiologist, after ensuring that he would not be quoted by name. "He's just so brilliant in so many ways, so he can sell his ideas. I don't think there are very many people who could sell that Aplysia model the way Eric has. But the question is, What is that ultimately going to tell us? Is it going to give us a wonderful model of Aplysia memory?"

And yet seeking basic biochemical mechanisms in simpler creatures is a time-honored method of biology. When evolution stumbles across a neat trick, it is likely to conserve it. DNA and RNA perform genetic roles in viruses and people. Surely a simple means to adjust the strength of synapses would be fruitful and multiply. To many neuroscientists, no one had done more to establish the biology of learning and memory than Kandel. And memory, according to his writ, was not postsynaptic, as in the theory of Gary Lynch, but presynaptic. In habituation, the flow of neurotransmitter is turned down, while in sensitization, the flow is turned up. It was the sending, not the receiving neuron that changed.

Kandel's early work was not necessarily incompatible with Lynch's. For years Kandel's detractors had been arguing that sensitization and habituation do not really qualify as learning. But in 1984, when Lynch published the calpain hypothesis, Kandel was working on a more ambitious theory that sought to explain classical conditioning, again using Aplysia. This theory of Pavlovian learning was also presynaptic. Kandel has been heard to praise Gary Lynch as a gifted, honest researcher. But he believed that with the calpain hypothesis Lynch had gone astray. It was no help to Lynch that a researcher with so much power to define the direction of neuroscience had come up with such a strikingly different solution. When a theory cooked up in a lab in the depths of Orange County, California, was stacked up against one emanating from Columbia University, many people assumed almost automatically that it was Kandel who must be right.

"I don't think people have any idea what his model is, you see," Lynch said one day as he reflected on the quirks of the scientific process. "I think it's just treated like one of the things that we have in neuroscience." Then, the typical qualification: "And, believe me, I'm not disparaging his hypothesis at all. I believe that Eric Kandel's body of work is a monument. It really is. If Eric Kandel were to walk out of neuroscience tomorrow, he would have left behind him a field. We obviously have our points of departure, Eric and I, but I do believe that in part his hypothesis is so well known because he's so well known."

. . .

Most people have an aversion to the words *classical conditioning*, as though the concept had been paired in their pasts with a painful stimulus like an electric shock. This, it seems, is the boring side of psychology, associated in people's minds with endless trials of punishment and reward, with Pavlov's dogs and B. F. Skinner's rats. In the Pavlov experiment, dogs learned that a bell was often followed by food, so they began salivating at the sound of the bell. In the jargon of behaviorism, the food is the unconditioned stimulus, the one that the dog responds to without training. The bell is the conditioned stimulus. Linking the two, through training, leads to some kind of connection in the animal's brain.

For most of the century, conditioning has been the province of psychologists, not biologists. Psychologists studied the phenomenon, developing precise mathematical theories about which training schedules led to the strongest links. Steeped in the behaviorism of Skinner, they rarely speculated about what might be happening inside the animal's brain. The whole point of behaviorism was to throw out concepts like mental representations and engrams. Since these mind objects could not be observed, Skinner believed, it was as silly to talk about them as it was to talk about the soul. Classical conditioning, on the other hand, was a behavior that could be observed and quantified. To the extremists like Skinner, all human behavior and even culture itself could be analyzed as stimuli and responses, input and output, a web of acquired behaviors that came to be known as S-R links.

Although such pure forms of behaviorism have fallen out of vogue, classical conditioning remains one of the most fundamental concepts in neuroscience. It is widely considered the most basic form of learning, the simplest case in which an organism learns to recognize a coincidence in its environment, to make a causal connection. Thunder means lightning. A flashing light means an electrical shock. One might argue that the very notion of cause and effect is rooted in classical conditioning. This sounds like an extremely reductionist stance, but Kandel would go it one better. The roots of causality might run deeper than psychology and even biology, he believed, all the way down to the level of chemistry.

Kandel was famous for brave reductionist leaps. The title of one of his papers was "Psychotherapy and the Single Synapse." In its simplest form, he suggested, classical conditioning occurred not at the level of the brain, where cells interacted with cells, but inside the neuron itself. It was, in other words, a molecular event.

This may not be as weird as it sounds. Respiration is, on one level, a macroscopic process. Lungs breathe and expel air. But on a microscopic level, respiration involves hemoglobin molecules, which absorb and expel oxygen atoms, changing shape like tiny lungs. And so it might be with classical conditioning. Traced to the bedrock, Kandel believed, the snapping together of a causal equation depended on the fact that in the cellular environment there are molecules capable of detecting the coincidence of two *biochemical* events. And it is these molecular reactions that stand for bells and meat, thunder and lightning. Learning in the outside world would be mirrored by molecular learning inside the cell.

. . .

From early in his career, Kandel was fascinated by the possibility of drawing connections between psychology and the behavior of neurons. After Harvard, he went to the New York University School of Medicine, where he studied psychiatry with every intention of becoming a psychoanalyst. But it soon became clear to him that the answers to his questions didn't lie at so high a level of analysis. He found his interest shifting to the biological substrates of thinking. For three years after graduation, he worked as a researcher for the National Institute of Mental Health in Bethesda, Maryland. Then, after moving to the Massachusetts Mental Health Center in Boston for his psychiatric residency, he went to France in the early 1960s to work with a neurophysiologist named Ladislav Tauc. It was there that he learned to experiment with the simple nervous systems of Aplysia, the "Philco radio" of the animal kingdom.

Probing with electrodes, Kandel and Tauc found in the Aplysia's neural wiring cells that seemed to receive inputs from two different pathways. This Y connection was just what Kandel had been looking for to perform a learning experiment. Was it possible, he wondered, that one of these inputs could be used as the conduit for a signal representing a conditioned stimulus, the other for an unconditioned stimulus? With proper training the neuron might learn that one stimulus was a predictor for the other: A pulse coming down the right-hand fork of the Y would mean that a signal was about to come down the left-hand fork. A change would occur in the cell that somehow linked the two events.

In a series of experiments, the scientists stimulated the first pathway with a signal of such low intensity that the neuron barely responded. Then, a few milliseconds later, they hit the second pathway with a strong signal, which would make the cell fire a neural spike. Most of the experiments didn't work. But occasionally, after a few trials in which the weak signal was followed by the strong one, the neuron seemed to make the connection. Now it would respond to the weak signal alone. A single cell seemed to have demonstrated the power to learn.

But it was one thing to find a physiological event that resembled classical conditioning in a nervous system and quite another to relate it to the animal's behavior. Somewhere in the brains of Pavlov's dogs, did a strong signal representing food converge with a weak signal representing a bell? Or had Kandel and Tauc been waylaid by an interesting biochemical artifact? The next step would be to train a whole snail to relate two stimuli, then see if learning caused changes in its nervous system.

As it turned out, years would pass before Kandel came close to carrying out this task. In 1963 he returned to the Massachusetts Mental Health Center, as a staff member, then two years later he transferred to his alma mater, the N.Y.U. School of Medicine. All the while he continued to study Aplysia.

No one expected this creature to be capable of very sophisticated behavior. Try as they might, Kandel and his colleagues, Irving Kupfermann, Vincent Castellucci, and Harold Pinkser, could not get Aplysia to demonstrate anything so sophisticated as classical conditioning. Still, the creature's nervous system was so simple and its cells so large and easily identifiable that the scientists were reluctant to give it up as a model. And so they concentrated on more rudimentary adaptive behavior, discovering the basic reflex that so much of Kandel's work would be based on: Tapping on the siphon causes the gill to withdraw.

Using a Water Pik to stimulate the Aplysia's siphon and a photocell to measure the amount of gill contraction, the scientists studied habituation, the tendency of an organism to respond less vigorously to a repeated stimulus. And in experiment after experiment, they painstakingly traced the neural circuitry involved in the gill-siphon reflex. There was no way to insert a probe into the tiny synaptic gap between neurons and actually measure a change in neurotransmitter flow. But using less direct techniques they were able to rule out other possibilities—that the receiving neuron became less sensitive, for example—and provide more and more circumstantial evidence for their theory. After four years, they were ready to make the case that habituation occurred when neurons in the gill-siphon circuit learned to produce less neurotransmitter. As a result, the sensory neurons in the siphon would send weaker signals to the motor neurons that caused the gill muscle to move. Shortly afterward, they showed that sensitization seemed to have the opposite effect: Repeated shocks to the tail made the gill-siphon reflex stronger because of an increase in neurotransmitter flow.

They still didn't know what was occurring in the neurons to cause the level of neurotransmitter to be turned up or down. But they began piecing together a theory. Recall that when an electrochemical spike travels to the end of an axon, where the vesicles of neurotransmitter lie, channels in the neuron open up, letting in calcium ions. And it is calcium that somehow causes the tiny balloons to attach themselves to the inside wall of the axon tip and burst, releasing neurotransmitter into the synaptic gap. Since Kandel's learning mechanism worked by regulating transmitter release, he assumed that it had to somehow make use of calcium. But what other chemicals were involved?

Off-the-Shelf Enzymes

AT THE TIME Kandel was formulating his hypothesis, it was becoming almost obligatory that any theory of neural functioning provide a starring role for a molecule called cyclic AMP. One might recollect from high school biology that the powerhouse of the body's cellular machinery is a molecule called adenosine triphosphate, or ATP, which has three phosphate groups, clusters containing a phosphorus atom. By breaking ATP down to ADP (with two phosphate groups) and AMP (with one), energy is released to power various cellular reactions. The phosphate bonds are like little batteries that store energy to be carried to the sites where it is needed. One of the breakdown products of ATP includes a ring-shaped structure—hence the name cyclic AMP.

In recent years scientists had learned that for some reason cyclic AMP was especially concentrated in the brain. Its purpose, it seemed, was to act as what biologists had come to call a secondary messenger, a transmitter of information that travels entirely within the confines of a neuron. Just as neurotransmitters carry messages from one neuron to another, secondary messengers carry information between the many parts inside a cell. As if neurons were not already complicated enough, scientists had discovered an entire communications network inside each one.

Cyclic AMP had been the subject of intense interest since the 1950s, when scientists were trying to figure out how hormones influenced other cells. Hormones can be thought of as neurotransmitters used for long-distance communication, from adrenal gland to legs, for example. When the brain wants to prepare muscles for sudden flight, it sends a signal by dumping adrenaline into the bloodstream. The adrenaline travels throughout the body, but only those cells with adrenaline receptors—muscle cells, for example—are able to respond. The adrenaline, it was found, doesn't actually enter the cells that it stimulates. Rather, it binds to receptors on the cell's surface. Then the receptor molecules relay the signal to other parts of the cell by using secondary messengers like cyclic AMP. By dispatching messages inside the cell's cytoplasm, the secondary messengers set off the biochemical changes—increased respiration, for example—that allow the muscle cells to work harder.

There is, in other words, a hierarchy of methods the body uses to communicate. To send signals from one creature to another, ants and other organisms emit pheromones, which are simply hormones that leave the body. While hormones and pheromones are used for biochemical broadcasting—the messages can be received by any cell equipped with the proper antenna—neurotransmitters are employed for more precise communication, within the synaptic gap. Finally, secondary messengers like cyclic AMP are used to communicate within the cell itself.

In the brain, the cyclic AMP machinery allows for a level of subtlety that could not be achieved by neurotransmitters alone. In the classical theory of neural signaling, receptors respond to neurotransmitters by opening ion channels, leading to the buildup of electrical charges that trips the cell's firing mechanism. By expanding the neuron's behavioral repertoire, secondary messengers allow for much more complex information processing. Neurotransmitters can be used not only to fire a neuron; under certain circumstances they can change it entirely by unleashing whole new reactions inside.

There are many ways this can happen. In some cases, hormone-like chemicals called neuromodulators bathe whole areas of the brain, activating special receptors that dispatch secondary messengers inside the cells. This sets off reactions that fine-tune the neurons, making them more or less likely to fire when they are stimulated by other neurotransmitters. Or, a neurotransmitter might unleash secondary messengers that tell the genes of a cell to manufacture a needed protein or produce an enzyme that will eliminate proteins that are already there. Since the very nature of a cell depends on the kind of proteins it manufactures, secondary messengers could conceivably cause long-lasting changes, the kinds of biochemical alterations that might be involved in memory.

In the case of cyclic AMP, this secondary messenger system was believed to work like this: A neurotransmitter, like serotonin, activates a special receptor and sets off a biochemical cascade. First, an enzyme called adenylate cyclase, which is somehow coupled to the receptor, starts making cyclic AMP. Then cyclic AMP activates another kind of enzyme called a protein kinase. This enzyme, in turn, activates still other enzymes. It does this by adding a phosphate group to them. This changes their shape, and proteins with different shapes perform different functions. Phosphorylating a protein can be thought of as a means of turning it on.

. . .

By the time Kandel was looking for a way to explain how a cell could raise and lower its own neurotransmitter flow, this cyclic AMP mechanism was sitting on the shelf begging to be incorporated into a learning theory. Including the cyclic AMP cascade in his hypothesis would not only save years of research. Any theory that can be woven into the preexisting fabric of neuroscience has a greater chance of being accepted as a solid contribution. Since more people will be familiar with the chemistry involved, the theory will be that much more credible. It would be wonderful for Kandel if sensitization occurred when a cyclic AMP chain reaction turned on enzymes that somehow resulted in more neurotransmitter release, perhaps by increasing the amount of calcium that came into a cell when it was firing.

In the late 1970s and early 1980s Kandel and his lab tested this idea with a number of experiments. In a live snail, sensitization occurred when a shock was delivered to the tail; from then on the animal would withdraw its gill more vigorously when its siphon was touched. In a dissected snail, Kandel and his colleagues performed what they considered a "neural correlate" of the experiment, stimulating nervous pathways with electrodes. Again think of a Y connection, a neuron receiving signals from two separate inputs. In this case a sensory neuron that fires when the siphon is touched also receives input from a second neuron carrying signals from the tail. In the experiment, a pulse through the siphon input (the right-hand fork) represented the tapping; a stronger pulse through the tail pathway (the left-hand fork)

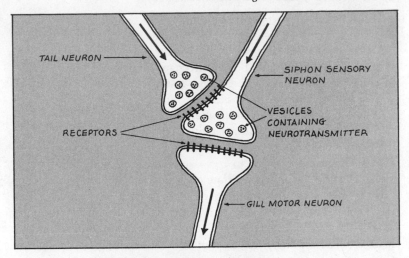

LEARNING IN A SEA SNAIL.
*During learning, the flow of neurotransmitter is turned up in the circuit
connecting the sensory neuron in the siphon with the neuron
that causes the gill to contract.*

represented the electrical shock. By jolting the tail pathway with an electrode, Kandel found he could produce what seemed like sensitization. Afterward, the siphon neuron would release more neurotransmitter in response to the tapping signal. This resulted in a stronger signal to the motor neurons that caused the gill muscle to withdraw.

Step by step, Kandel simplified the experiment. First, he discovered that he could eliminate the shock signal entirely, sensitizing neurons by stimulating them with a dose of serotonin, the chemical he believed to be the neurotransmitter involved in the gill-reflex circuits. But what if he went another step and simply injected cyclic AMP directly into the cell? When the researchers tried this they found that sensitization still occurred. Then they repeated the experiment, eliminating yet another middleman. They injected kinases, the next step in the biochemical chain reaction, into the cell and showed that this chemical alone also caused a sustained increase in transmitter flow. The kinases presumably were turning on enzymes by phosphorylating them, but what proteins were being activated this way?

Taking a cue from Paul Greengard of Rockefeller University, Kandel and his colleagues speculated that, once activated by cyclic AMP, a kinase phosphorylates some of the proteins that act as the neuron's potassium channels, the portals that control the flow of potassium ions in and out of the cell. Potassium channels play an important role in neural transmission. During firing, sodium ions rush into the neuron and potassium ions rush out.

Kandel proposed that in its altered, phosphorylated state, a certain type of potassium channel would be stuck closed. Now, when the neuron fired, the potassium couldn't move out of the cell as easily as before; there would be an increase in the duration of the neural spike. Held in this prolonged state of arousal, the neuron would have time to take in more calcium. And it is calcium that activates the mechanism that leads to neurotransmitter release. Sensitization, then, altered the chemistry of the cell. The result was a neuron that now reacted more strongly to a stimulus by emitting more neurotransmitter.

. . .

Having proposed a biochemical explanation for sensitization, Kandel was ready to take on classical conditioning. By 1981, he and his colleagues had finally got Aplysia to demonstrate what they claimed was a simple form of conditioning, although the interpretation of the experiment is still hotly disputed. In the earlier experiments the electrical shock to the tail was used to sensitize the snail, causing it to respond more strongly to the tapping on its siphon by withdrawing its gill. But as they played around with Aplysia, trying to uncover other reflexes to study, Kandel and his colleagues found that a tail shock alone could also cause the gill to withdraw. In the new experiments they discovered that if they repeatedly followed a touch of the siphon with a shock to the tail, they could get Aplysia to learn that the touch was a predictor for the shock. Here the electric shock was the unconditioned stimulus. Without any training, it caused the Aplysia to reflexively withdraw its gill. A light tap to the siphon was the conditioned stimulus. At first it caused only a slight gill withdrawal. But if the scientists paired the two stimuli, always following the tap by the shock, the snail would learn to vigorously withdraw the gill just at the tapping.

But had the snail really formed an association, or was this just another example of sensitization? Perhaps the shock simply aroused the snail so it was hypersensitive to the tapping. Aplysia's nervous system was so crude that it sometimes seemed like one big short circuit. The difference between sensitization and what Kandel accepted as classical conditioning seemed pretty subtle, as his rivals were quick to point out. Some considered the distinction imaginary. But Kandel insisted there was a difference. Rapidly pairing the stimuli—tap, shock, tap, shock—caused a stronger gill withdrawal, he said, than resulted from sensitization alone. If the experimenters waited more than a second between the two stimuli, the snail wouldn't learn the response as well. Nor, as is usual in classical conditioning, did Aplysia seem to make the association if the two stimuli occurred simultaneously or if the order was reversed. In Pavlovian terms, presenting food before the bell doesn't result in learning. As Kandel explained it, the siphon neuron—the junction of the Y connection—acted as a kind of coincidence detector. Its neurotransmitter flow was turned up to its highest level when the cell received first one signal, and then the other within a narrow window of time.

Depending on whom you talk to, the biochemical explanation for all this is either a milestone in the resolution of the mind-body problem or one of Kandel's more skillful acts of intellectual salesmanship. Again the distinction between sensitization and conditioning is so subtle that trying to untangle the two biochemical pathways can be a dizzying experience.

According to his theory of conditioning, when the siphon neuron receives the tapping signal, it responds as it would to any stimulation. It fires, as various ion channels open, including ones that let in calcium, the trigger for neurotransmitter release. Then, when the neuron receives the signal from the tail neuron, indicating an electric shock, the cyclic AMP mechanism is turned on, the same one that in the earlier experiments just led to sensitization. Here is the difference: In classical conditioning, the shock signal comes so quickly after the tapping that the neuron still has excess calcium inside. This calcium, Kandel believed, somehow amplifies the cyclic AMP cascade. As in sensitization, cyclic AMP uses kinases to phosphorylate potassium channels, causing them to stay closed. And this prolongs the cell's state of arousal: From now on, every time it fires it lets in more calcium than it used to, and this leads to greater neurotransmitter release. In conditioning the effect is simply more pronounced. With the cell altered to become a more copious producer of neurotransmitter, all it takes in the future is the light tap on the siphon to cause the same vigorous reaction that once required an electrical shock. This effect soon fades, but in long-term memory, cyclic AMP might activate other proteins. These could include ones that act as signals to the genes, telling them to make fewer potassium channels or more calcium channels. Then the cell would be radically altered. Neurotransmitter flow would be permanently turned up.

Inside the cell, then, the chemicals would act as symbols. Calcium stands for tapping, cyclic AMP for the electric shock. It is still not clear, however, how the symbols become linked. The implication is that somewhere in the cell there is a single enzyme that connects these two chemical events.

The coup de grace to Kandel's conditioning theory came when he proposed that this coincidence-detecting molecule is adenylate cyclase. This is the enzyme that forms the first step in the cascade, responding to the neurotransmitter from the tail-shock signal by making cyclic AMP. According to the theory, the molecule has a binding site for calcium. When it detects the presence of both neurotransmitter at the receptor and calcium in the cell, it is much more effective at setting off the reactions that ultimately lead to channel phosphorylation and increased neurotransmitter flow.

If Kandel was right, his theory would represent what he called "a surprisingly radical reductionist possibility." Our very notion of causality could be rooted in the ability of certain molecules to make an association between two chemicals that stand for two signals that stand for two events in the outside world.

Religious
Feuds

THROUGHOUT THE SECOND HALF of the 1980s, Kandel continued to work on his theory and promote it at neuroscience meetings. It was a neat story, though not as easy to visualize as the one told by Gary Lynch. It seemed, for example, that the action of the adenylate cyclase molecule would be turned up whether the calcium came before or after the neurotransmitter. Why then did the order of the stimuli matter, as it must in true conditioning?

More damning, perhaps, was that Kandel's theory failed what many memory researchers considered the acid test: It was not "specific to the synapses." Learning caused the whole cell, not just a single synapse, to change. In his landmark paper, Lynch called the calpain mechanism "A New and *Specific* Hypothesis." With the subtlety customary in scientific publications, he was thumbing his nose at Presynapticists like Kandel. In Lynch's postsynaptic theory, LTP unleashed the calpain mechanism only in the synapses that had been stimulated; they became more sensitive while the other synapses remained the same. If, as in Kandel's hypothesis, learning caused the sending cell to release more neurotransmitter, then it seemed that every synapse it fed into would be turned up.

"It's like throwing a hand grenade into the cell," said James Olds, a researcher at the National Institute of Mental Health. "The whole thing flips."

It was hard to see how this crude change could allow for the fine-tuning necessary to etch detailed circuitry. Most scientists believe that a single neuron can participate in any number of memory patterns— it becomes connected through its thousands of synapses into thousands of overlapping neural circuits. Think again of the letters and numbers on an electronic scoreboard—a single bulb participates in many different patterns. But if the whole cell, and not just the synapse, is the unit of change, then a bulb used for one engram would be unavailable for any others. Once a cell had been altered it would be used up, along with all the synapses it fed into. There are billions of neurons but trillions of synapses. Kandel's theory provided for a vastly smaller memory store.

Kandel could conceivably add some gears to his mechanism that would make it specific to the synapses. But that was the least of his troubles. There had always been those who complained that the Aplysia story was just not as neat as Kandel would have it. The main objection was that his laboratory used different animals for the behavioral experiments and the physiological experiments. He had shown that he could train Aplysia to modify the gill-withdrawal reflex. And he had shown that, in a dish, the neurons in the gill-withdrawal circuit could be modified by stimulating them with electrodes. But he had never trained an Aplysia, cut it open, and shown that learning actually caused the cellular changes.

Most neuroscientists seemed to accept the value of his work on habituation and sensitization. But throughout the 1980s his theory of classical conditioning was coming under increasing attack. His main nemesis was Daniel Alkon, a researcher at the National Institute of Mental Health, who insisted that what Kandel had done with his snails "cannot even remotely be considered classical conditioning or associative learning."

Classical conditioning, Alkon argued, must involve the acquisition of an entirely new skill. In his experiments, Kandel was simply amplifying a response that was already there. Untrained snails plucked from the sea withdraw their gills when their siphons are tapped. Sensitized snails, whose tails have been shocked, respond to the tap more vigorously. Conditioned snails, for which the tap was paired with the shock, respond more vigorously still. But nothing new has been learned. As Alkon once put it, exaggerating only slightly, "The control group and the experimental group are the same." The difference was all a matter of degree. Before training, Pavlov's dogs didn't salivate to bells at all.

At his laboratories in Bethesda, Maryland, and Woods Hole, Massachusetts, Alkon worked with another kind of snail called Hermissenda. Among this creature's rather limited behavioral repertoire is the tendency to move toward light (in search of food) and to respond to turbulence by contracting its foot, holding tight to the bottom of the sea. Using an elaborate apparatus consisting of a turntable, glass tubes, and photoelectric cells, Alkon trained the Hermissenda to associate light with turbulence. Shine a light on a trained Hermissenda, and it would steady itself by contracting its foot, as though it were anticipating a bad storm. An entirely new behavior had been learned.

This, Alkon believed, was much less artificial than the procedure Kandel followed. "In nature, electrical shocks don't occur," Alkon said. "Using shocks you might induce changes that look like learning, but the chances that you are tapping into a natural system are very low." By studying a natural response—foot contraction caused by turbulence—Alkon believed he was working with real circuits. He traced the wiring for the reflex (the blueprint, he called it) and used the same animals for both the behavioral and the physiological experiments.

In Alkon's theory of conditioning the main actor was an enzyme called protein kinase C. Through a complex series of chemical reactions, conditioning caused the enzyme to move from the cell body to the dendrite. In this new location the enzyme phosphorylated potassium channels, blocking them by sticking on a phosphate group. But here Alkon's and Kandel's theories radically diverged. In Kandel's scenario, the blocking of potassium channels occurred in a sending cell, causing it to emit more neurotransmitter. In Alkon's theory, the changes occurred in a receiving neuron, making it more sensitive to stimulation. After repeatedly pairing a light and rotation, the cell would become sensitive enough to respond to light alone, sending a signal that caused the animal to steady itself with its foot. Like his rivals, Alkon believed that, over time, training also would lead to chemical reactions that talked to the genes, telling them to synthesize the proteins needed for long-lasting structural change.

As in Lynch's calpain hypothesis, the kinase C mechanism was postsynaptic, not presynaptic. But Alkon quickly ran into the same problem that Kandel did: It was difficult to explain how the changes would be limited to a single synapse. Why wouldn't a whole cell and all its synapses become more sensitive? Alkon was convinced that his mechanism was somehow specific to the synapses. The kinase, he believed, was transported only to the areas of the dendrite where incoming signals needed to be amplified. The rest of the synapses would remain available for use in other circuits. But he knew that he was a long way from explaining just how the kinase would know where to go.

In the meantime he tried to show that he was not guilty of what so many researchers accused Kandel of doing: spending a career elaborating an obscure mechanism in sea snails that might have nothing to do with human learning. Alkon's lab also experimented with rabbits. Blow on a rabbit's eye, and it will blink; pair the puff with an auditory tone, and the animal will learn to blink at the sound. By dissecting the hippocampuses of trained rabbits, Alkon and his colleagues believed they had found evidence that learning activates the same kinase C machinery. In fact they showed that in both Hermissenda and rabbits, potassium currents were reduced for days and even weeks after training. In one dramatic experiment, Jim Olds used the tone-puff procedure to train a rabbit. When the lesson was over, he cut up its brain. Using a radioactive label, he found that kinase C seemed to have migrated from the cell body to the dendrites, just as Alkon had predicted.

Unfortunately for Alkon, he and other researchers also found that training a rabbit to associate a tone and a puff can cause as many as half of the neurons in one region of the hippocampus to chemically change. It seemed that it was not just the whole cell that flipped, but a large chunk of the brain. Again, Alkon hoped that he could come up with some kind of "smart" transport mechanism that would explain how, once kinase C production had been switched on in a whole population of cells, the molecule would only turn up the volume of the appropriate synapses.

· · ·

And so the debate between the Presynapticists and the Postsynapticists continued. Wars have been fought over subtler distinctions. The success of Kandel's theory would not be fatal to the calpain hypothesis. After all, he was looking at procedural, not declarative learning, the learning of skills, not facts. More threatening to Lynch's theory was Alkon's mechanism. He and other scientists, like Aryeh Routtenberg at Northwestern University, had made some headway in showing that the kinase-C machinery is also turned on by LTP. In all these theories, calpain—if it plays any role—was merely part of the background noise.

"My theory is less flashy than Gary's," Routtenberg conceded, but he, too, believed the calpain hypothesis was wrong. For one thing, he was not convinced that calpain actually existed inside the synaptic spines. Even if it did, he could show evidence that calpain simply helped regulate the action of kinase C. So any evidence Lynch provided for his mechanism—that it was disrupted by calcium chelators or by the calpain blocker leupeptin—Routtenberg could use to support his own hypothesis. Each scientist could subsume the other's theory as a detail.

The fact that the same elements showed up in so many different learning theories might suggest that the scientists were all on the right track, converging on the same target from different directions. But that was not necessarily the case. A neuron contains thousands of different kinds of proteins and enzymes, which are involved in an enormous number of reactions: breaking down sugar to release energy, synthesizing new proteins by reading the genetic blueprint and sending the instructions to the ribosomes, transporting proteins to the proper places in the cell. Calcium and cyclic AMP are used as secondary messengers in any number of circuits of internal communication. Protein phosphorylation is regularly used to turn on enzymes. At any point in the history of a science, there are only certain colors on the palette, and they go in and out of style.

In fact, cells might very well have evolved a number of different learning mechanisms, some for short-term memory, others for long-term; some for storing facts in a network, others for instilling new behaviors in a single cell. With so many possibilities it is easy to come up with compelling configurations. In the end, however, most of them may well turn out to be wrong.

"A Beautiful Little Switch"

SEVERAL YEARS after the publication of the calpain hypothesis, Lynch and Baudry encountered an unpleasant surprise. Some of the most important evidence for the theory still came from the test tube experiments measuring what they took to be an increase in glutamate binding—a sign that LTP caused new receptors to appear. Of course they would have preferred not to rely on such indirect arguments, but none of the advances in neurobiological techniques had removed a fundamental obstacle: It was impossible to see firsthand that LTP actually broke the cytoskeleton and caused hidden receptors to pop out. Measuring glutamate binding seemed the next best thing. In experiment after experiment they took tissue that had undergone LTP, liquefied it, added glutamate to the mixture, and showed that the residue—consisting, they believed, of neurotransmitter molecules latched tightly to glutamate receptors—had increased.

"For four or five years the whole story was just beautiful," Baudry said. "Everything was so consistent. Then things started getting more complicated, as usual."

Experiments from another laboratory showed that it was not glutamate binding that was occurring in their test tubes after all, but some other chemical process. In fact they were no longer sure exactly what they had been measuring. "This whole part of the story is now a question mark," Baudry lamented. This didn't disprove the hypothesis that LTP turns up the synapse by increasing the number of glutamate receptors, but it left Lynch and Baudry with a serious hole in their theory.

"I rue the day we ever said those were receptors," Lynch said. "It was a tactical error of the first magnitude."

But it is hard to hit a moving target. Science is not simply a matter of collecting facts. The arrangement is everything. One builds a structure of inferences, snapping it together piece by piece. When a model clashes with reality, as revealed by an unwelcome but undeniable experimental result or an internal inconsistency in the theory, the builder must make adjustments. The difficulty is knowing when to give up. Theories, like bureaucracies, can take on lives of their own. Sometimes a scientist will do anything to preserve his creation.

In an attempt to save the calpain hypothesis, Lynch and Baudry began speculating on other ways a memory trace might be left without requiring the appearance of new glutamate receptors. In his original electron micrograph experiment, Lynch had seen evidence not only that LTP caused new shaft synapses to form but that the preexisting synapses were undergoing a subtle change. Their spines—the dendritic bumps on which synapses are usually made—seemed to become stubbier and rounder. Since the role of calpain was to break the cellular skeleton, it might be responsible for this change. LTP would unleash calpain, calpain would break the cytoskeleton, and the spine would change shape. Rounder spines, Lynch speculated, might be more excitable, perhaps because of a change in their electrical resistance. Then, even with a constant number of receptors, a synapse that had been changed by learning would react more strongly to neurotransmitter; it would be more likely to trip the mechanism that causes the cell to fire.

But other scientists remained dubious. Greenough's research suggested that what Lynch believed was a rounding of dendritic spines was really an artifact, caused not by LTP but by the chemicals he used to preserve the brain slices.

Lynch didn't much like spine rounding either—it was just not as elegant as that notion of snapping the cytoskeleton so hidden glutamate receptors could emerge. This latest speculation was more of a holding action, allowing Lynch to try to establish what he considered the most important part of his hypothesis: that calpain can actually cause entirely new synapses to appear.

And for all the difficulties with his theory, Lynch could take comfort in one fact: It remained the only one that was clearly specific to the synapses. He didn't need to posit all kinds of complex, possibly imaginary machinery to explain how, for example, kinase C was transported to the proper synapses. Neurotransmitter turned on receptors, calcium came in through the membrane and activated calpain, calpain ate the cytoskeleton and caused the synapse somehow to change. All the action took place right there in the synaptic spine, just where it should.

. . .

During the next few years, Lynch's laboratory concentrated on refining its knowledge of LTP and how it might lead to the creation of memory circuits. An important clue had come back in 1978 when Bruce McNaughton at the University of Colorado showed that LTP is especially effective when not one but several pathways leading to a group of neurons are stimulated simultaneously. For scientists trying to find links between psychology and biology, this was very suggestive: One can imagine signals from several circuits colliding at a single neuron. This convergence somehow triggers LTP. The synaptic connections are strengthened, and the circuits are welded together. One signal might come from a structure of neurons that somehow represents the taste of an apple; the second signal from a structure that stands for a feeling of roundness; the third from a structure that symbolizes the color red. Piece by piece, the brain would string together an associative memory called apple.

But how does a synapse know when to potentiate, making a circuit that didn't exist before? Obviously, if LTP occurred every time a synapse was stimulated, everything in the brain would quickly become connected to everything else, resulting in a massive cerebral breakdown.

In the years following the publication of the calpain hypothesis, John Larson, a postdoctoral student in Lynch's lab, carried out some now classic experiments with rats that suggested what kind of rules are followed when the neural wiring is laid in place. Larson discovered that potentiation is a two-step event. The first signal, consisting of a short burst of high-frequency pulses, travels in through one input and primes the neuron; the second signal, coming down a different path, causes LTP. Timing, Larson found, is everything. The second signal is most likely to cause LTP if it follows the priming signal by about two hundred milliseconds. The first signal seemed to tell the neuron, "Okay, if you receive another signal within the next fifth of a second, make a link with the neuron that sent it. Otherwise ignore it."

Larson found that synaptic strengthening was especially effective if he repeatedly zapped the first input then the second, back and forth, always maintaining the two-hundred-millisecond interval. The result was a frequency—five hertz (the number of cycles per second) —that corresponded to the so-called theta rhythm, one of the brain wave patterns that can be measured with an electroencephalograph. It was exciting enough that Larson had linked LTP to a natural brain rhythm. Even more suggestive was the fact that the theta rhythm has been shown to emanate from the hippocampus when a rat is exploring its surroundings, gathering information. Other scientists have proposed that the hippocampus is where the rat stores its mental map. If electrodes are put in various cells of the hippocampus, they each fire like Geiger counters when the rat approaches a certain part of a maze. If Larson and Lynch were correct, the theta rhythm was a sign that the rat's cerebral computer was filing away information, making internal maps by using calpain to form new brain circuitry.

After the disappointment over glutamate binding, Larson's findings were a godsend. "We have found the magic rhythm that makes LTP," Lynch said, one day shortly after this latest development. "There's a magic rhythm, the theta rhythm, the natural, indigenous rhythm of the hippocampus."

But what chemistry could account for the theta rhythm? The answer to that question helped nudge the calpain hypothesis back toward the mainstream of neuroscience.

.   .   .

During the last two decades, scientists had become increasingly convinced that there were not one but several kinds of glutamate receptors. At first this seemed more an annoyance than a revelation. Once again the waters had been muddied. But as the debris began to clear, one of these molecules, the so-called NMDA receptor, began to emerge as a vital component in theories of how memory works in the mammalian brain.

The name of the receptor is not particularly important—NMDA is simply a chemical, N-methyl D-aspartate, which is used to identify it in experiments. Although NMDA itself is not known to exist in the brain, it is similar in structure to glutamate. It is what neurochemists call an agonist, a molecule that, like a neurotransmitter, stimulates a receptor. The opposite of an agonist is an antagonist, which interferes with the receptor, probably by blocking its binding site, preventing the neurotransmitter from latching on. By juggling agonists and antagonists, experimenters were able by the early 1980s to sort out several kinds of glutamate receptors, each with different properties.

At first, these findings were of no real interest to memory researchers. Then, in 1983, a British scientist named Graham Collingridge showed that LTP will not occur if a neuron's NMDA receptors are turned off. Using an antagonist called APV (aminophosphonovaleric acid), he blocked the receptors in a sample of brain tissue, then stimulated a pathway with high-frequency pulses. For the most part, the neurons seemed to work just fine; cells would fire and send their glutamate signals, which were received as usual on the other side of the synaptic gaps. But while the receptors involved in normal transmission were unimpeded, LTP would not occur.

Apparently, there was something special about NMDA receptors that was integral to memory—or at least to LTP. Researchers soon discovered that the receptor was unlike any other. Receptors generally work by opening ion channels in the cellular membrane. Until very recently, all ion channels were believed to be either chemically gated or electrically gated. Either the arrival of a neurotransmitter at the receptor or a change in membrane voltage caused them to open and close.

The NMDA receptor had the distinction of being both electrically gated and chemically gated—it required both kinds of stimulation before it would open its ion channel. Normally, a magnesium ion blocked the channel. No matter how much neurotransmitter crossed the synapse and stuck to a neuron's NMDA receptors, they would be incapable of responding. But if the neuron was already in a state of electrical arousal—stimulated by a previous signal that had activated the cell's normal glutamate receptors—then the magnesium stoppers would pop out. Now the NMDA receptors were free to react to a second rush of glutamate by opening their channels. The receptor was, in other words, a two-step device. One pulse cocked the trigger, the second pulse fired the gun. And while normal receptors worked by allowing sodium ions into the cell—the positive charges that led

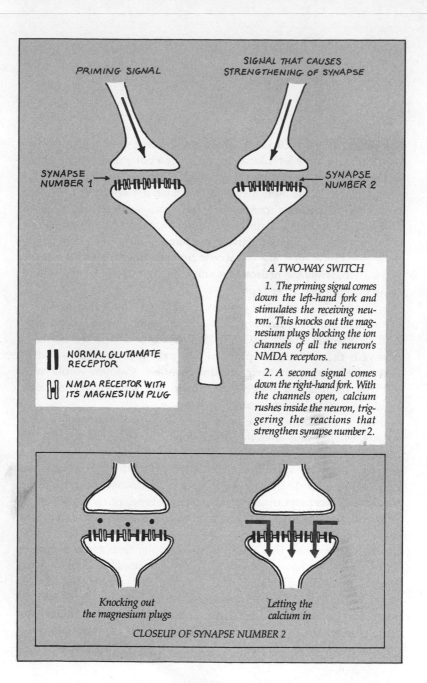

PRIMING SIGNAL

SIGNAL THAT CAUSES
STRENGTHENING OF SYNAPSE

SYNAPSE
NUMBER 1

SYNAPSE
NUMBER 2

A TWO-WAY SWITCH

1. The priming signal comes
down the left-hand fork and
stimulates the receiving neuron. This knocks out the magnesium plugs blocking the ion
channels of all the neuron's
NMDA receptors.

2. A second signal comes
down the right-hand fork. With
the channels open, calcium
rushes inside the neuron, trig-
gering the reactions that
strengthen synapse number 2.

‖ NORMAL GLUTAMATE
RECEPTOR

Ж NMDA RECEPTOR WITH
ITS MAGNESIUM PLUG

Knocking out
the magnesium plugs

Letting the
calcium in

CLOSEUP OF SYNAPSE NUMBER 2

FORMING A MEMORY CIRCUIT.

to the firing of an action potential—the NMDA receptor also let in calcium.

· · ·

When Gary Lynch heard about the existence of a glutamate receptor that not only let calcium into a neuron but worked in a manner similar to John Larson's two-way switch, he could hardly believe his good fortune. In his mind's eye, he imagined a Y-shaped circuit with two sending neurons converging on a single receiving neuron. Through the left-hand pathway a signal would discharge glutamate into the synaptic gap and activate the target neuron's normal, chemically gated receptors. The cell would fire, sending its spike down the axon. Then, while the neuron was still aroused, a second signal would come through the other input.

This time, all over the neuron, NMDA receptors that had been quiescent during the first signal would be primed for a second one. And so it would come, down the right-hand fork. With the blockage removed, the NMDA receptors in this pathway would respond to glutamate by opening their calcium channels. Calcium would rush into the neuron and trigger the calpain mechanism. The synapse would be strengthened, a connection made.

The signals didn't have to come down separate pathways. Two signals arriving one after the other down the left-hand fork would cause that synapse to change. In fact, it didn't matter which of a neuron's many synapses the second signal came through. Any signal that arrived while the target neuron was still aroused would activate the NMDA receptors in the appropriate synapse, turning on the calpain mechanism—or some set of chemical reactions that required calcium as a trigger.

Lynch assumed that this was the two-step process that Larson had observed in his experiments. To test the possibility, he and his student infused brain slices with APV, the drug that blocks NMDA receptors, and showed that it interfered with LTP; without the NMDA receptors working, the theta rhythm no longer caused as much synaptic change. Lynch was exultant. It seemed that, step by step, they were converging on the prey.

One day, shortly after the experiment, Lynch sat in his office. Primed with this new evidence, he was pushing the calpain hypothesis as vigorously as ever. One by one, he ticked off the arguments in his favor. "We know a brain rhythm that will activate a receptor that will pass calcium into a cell. We know that calcium activates a protease that tears apart a cytoskeleton protein. We know that if you were to do that in a blood cell, you'd change the living hell out of it."

There were still many missing links in the chain. The calcium let in by the NMDA receptors could be acting as a trigger for any number of cellular reactions, including those of Alkon or Kandel. Indeed, some experiments have shown that activating NMDA receptors leads to a rise in the level of protein kinase C, one of the enzymes that phosphorylate ion channels—the key mechanism championed by Lynch's rivals. Everyone, it seemed, was trying to rope NMDA receptors into their theories. If nothing else, linking the calpain mechanism to this strange molecule was a good tactical move.

"See, one of my problems in life is that my interpretations are never sexy," Lynch said. "My interpretations are rarely au courant. I never explain anything with protein phosphorylation. I rarely invoke cyclic AMP."

But like protein phosphorylation, the NMDA receptor was rapidly becoming one of the hottest areas of neuroscientific research. The excitement might turn out to be wrongheaded. The reason so much is known about the receptor is simply that scientists happen to have a good agonist and antagonist with which to study it. Sitting next to it might be an even more important receptor, invisible for lack of the proper chemical tools. But throughout the 1980s and into the 1990s, there has been an overwhelming sense that this newfound molecule is going to turn out to be the real thing, a way to launch new careers, revitalize old ones, and maybe even contribute something important to the understanding of memory.

One of the first researchers to leap into the fray was Carl Cotman, an Irvine neurobiologist who collaborated years ago on Lynch's sprouting experiment. "This receptor is the first concrete property of a synapse that allows you to go from a detailed molecular mechanism all the way up to learning and memory and to higher cognitive functions," he said one day in 1988. "It's a beautiful little switch."

Hebb's
Return

PART OF THE EXCITEMENT caused by the discovery of the NMDA receptor was that it bore such a striking resemblance to the Hebb synapse: When two neurons tend to fire simultaneously, the connection between them is strengthened. When Hebb made this suggestion in 1949, it seemed to make such good sense that scientists had been looking for evidence ever since. The situation was not unlike the hunt for the quark, though decidedly less intense. When the physicist Murray Gell-Mann proposed these hypothetical particles, they provided such an elegant means of classifying the seeming chaos of subatomic particles that experimenters quickly began searching for evidence of their existence. Now they are accepted as phenomena, not just mathematical conveniences.

The Hebb synapse was held in such esteem that any theory that invoked it almost automatically gained an air of respectability. Throughout the 1980s, a number of researchers, including Bengt Gustafsson and Holger Wigström in Sweden and Thomas Brown of the City of Hope Research Institute in Duarte, California, showed in experiment after experiment that LTP was strongest when the sending and receiving neurons were active at about the same time. Though this sounded suspiciously like Hebb's synapse, it wasn't until the discovery of NMDA receptors that scientists had a molecule that seemed to explain how such a mechanism might work. When Hebb put forth his theory in 1949, scientists hadn't an inkling there was such a thing as LTP, much less a chemical switch called the NMDA receptor. Experiment seemed to be catching up with theory.

· · ·

In 1986 a team of British scientists joined with Lynch and Baudry to make the strongest link yet between the NMDA receptor, LTP, and memory. After some discussion with the Irvine scientists, a Scottish researcher named Richard Morris came up with an ingenious way to show that chemically blocking NMDA receptors interfered not just with LTP but with learning.

In the experiment, rats were dropped into a small swimming tank. To keep from drowning they had to learn to find a platform to climb upon. At first they would swim at random, all over the tank, until they chanced upon the platform. But after several trials they would learn to head immediately for the proper corner, circling until they converged on the target.

Then Morris repeated the experiment on rats whose NMDA receptors had been blocked by APV, the same drug Collingridge had used in his LTP experiment. The rats' trajectories were recorded by an image-sensing device. After several learning trials, normal rats would quickly lock into a swimming pattern that zeroed in on the platform. But with rats whose brains had been infused with APV, the convergence never occurred. They continued to swim randomly around the tank. Without NMDA receptors firing, all the experience had apparently been for naught. The engrams were never laid.

But the scientists found that other kinds of learning were not blocked by APV. In another experiment, two platforms were put in the tank. One was secure; the other would sink if a rat tried to climb on it. To distinguish the platforms, one was marked with black and white stripes, the other was gray. In this case, both groups of rats— the normal ones and the ones with blocked NMDA receptors—learned the task with about the same ease. The receptors, it seemed, were involved in only certain kinds of learning. In the first experiment the rats had to make a mental map of the swimming tank. In the second experiment they simply learned to tell the difference between stripes and solid gray. The distinction between these two forms of learning is not exactly clear-cut. But Lynch took it as further proof that NMDA receptors—and, he hoped, the calpain mechanism—were involved in declarative learning, the memorizing of facts, not procedures.

· · ·

Not all the evidence for the new theories came from experiments with rats. Anatomists working with human brains also found links between NMDA receptors and memory. The most striking case was that of R.B., a retired Southern California postal worker who, in the aftermath of a coronary bypass operation, suffered a sudden loss of blood to his brain. Though he survived with most of his faculties intact, he lost his ability to remember. He could recall events that had happened in the years before the operation, but he could not form new memories.

When R.B. died in 1983, his last half-decade an amnesic blur, three scientists were allowed to dissect and study his brain: Stuart Zola-Morgan and his colleague Larry Squire of the Veterans Administration Medical Center of San Diego, and David Amaral of the Salk Institute. R.B.'s amnesia had not been caused by massive or even moderate brain damage, they discovered, but by a tiny lesion in the hippocampus. Most important, perhaps, is that the brain cells that had been damaged are now known to be unusually rich in NMDA receptors.

With so many developments coming so fast, it was hard to sort them out. Was this receptor really so amazing, or was the flood of data simply a function of the sheer number of researchers ganging up on the same molecule? The closer they looked, scientists seemed to find it capable of the most subtle interactions. Recent evidence indicated, for example, that the receptor is more sensitive when it is exposed to an amino acid called glycine.

"Now you've potentially got a way to turn it up and turn it down," Carl Cotman said. It seems the receptor not only acts as a two-way switch but is equipped with a volume control as well.

Intuition and Ambiguity

As THE 1980s drew to a close, scientists were still debating whether the mechanisms of memory were presynaptic or postsynaptic. At the 1988 Society for Neuroscience conference in Toronto, Lynch and some colleagues reported that they had linked NMDA receptors to the breakdown of spectrin, the cytoskeletal protein. But then everyone with a learning theory was trying to come up with ways to get it to work with NMDA receptors. For the Presynapticists this was especially awkward. After all, the receptors were active in the receiving, not the transmitting neuron. How could they cause the sending neuron to change?

Where there's a will, there's a way. During one session of the conference, Timothy Bliss, the co-discoverer of LTP, stood among the dozens of posters describing the latest developments in learning and memory theory and nervously explained why he himself remained a staunch Presynapticist: Pointing to the pictures and charts on his own poster, he showed how the activation of the NMDA receptors might cause the receiving neurons to send a "retrograde" messenger back across the synapse to the sending neuron, turning up transmitter flow.

But to many of his rivals, the mechanism seemed rather labored and inelegant, a heroic stretch to make NMDA receptors consistent with the tenets of Presynapticism. At the same conference, Lynch's laboratory reported findings that they believed ruled out, once and for all, the possibility that LTP was presynaptic. Late one night in the bar at the Toronto Sheraton, he and his friends celebrated what he took to be the demise of the rival creed. "This is a historic moment in biology," he said, with a bit of irony, as he exultantly anticipated what he hoped would soon be the coup de grace: an experiment that would blow protein kinase C and cyclic AMP out of the water, taking with it Kandel, Alkon, Routtenberg, Bliss—whole schools of enemy learning theories.

But for all the rhetoric, none of the memory mechanisms is ready to be dignified by the word *theory*. Some scientists have suggested that the changes caused by learning might be both presynaptic and postsynaptic; that would surely seem reasonable if it turns out that synapses are not only strengthened but also created from scratch. In the end, it is only the winning hypotheses that are remembered, and it's possible that Lynch might be building, inference by inference, an elaborate, elegant, meaningless architecture, a tower to nowhere.

"We could be wrong," he said one afternoon in his office, fingering a cigar. On the bookshelves behind him, models of primitive human skulls looked over his shoulders. Tape cassettes were scattered around the room. "This hypothesis about calpain and all the rest of it could be dead wrong. There will be tons of good cell biology that came out of it. But as a memory mechanism this could be wrong. There goes eight years.

"What you're really banking on is your intuition," he said. "I mean, you construct these stories so that they all make sense; you go through the arguments for what you feel the truth is. Why are you doing this? You're doing this because of some strong intuitive urge that says, Yeah, this feels right."

Lynch is still waiting for the experiment—perhaps there is none—that will show the laying down of engrams as clearly as Du Bois-Reymond demonstrated the action potential. One thing that might provide such solid evidence would be a really good calpain blocker; for a number of reasons, leupeptin doesn't produce conclusive results.

"This whole thing could be confirmed or come crashing down around our ears when somebody comes up with a really good calpain inhibitor," Lynch said. "Now at that point, we might shoot it into a rat and get a really dramatic amnesia." Or else a rat whose memory still works just fine.

Unlike protein phosphorylation and NMDA receptors, calcium-activated proteases are still an obscure corner of brain research. Ironically, they keep popping up in theories that explain not learning but brain damage. During epileptic seizures or conditions of blood loss—like the one suffered by R.B.—neurons fire wildly, secreting a flood of glutamate. As during LTP, the NMDA receptors respond by letting calcium, in this case a deluge of calcium, into the cells. Could it be that calcium activates proteases in so great a number that they begin to eat up brain cells? Perhaps it wasn't just a coincidence that of all the cells in R.B.'s brain, it was the ones that were especially rich in NMDA receptors that were destroyed. It may have been the receptors themselves that triggered the damage. Memory then might contain its own self-destruct mechanism. Just as an acid can be used to engrave the finest of etchings, so can it destroy its own creation. Brain damage, it seems, might be a case of learning gone wild.

To hear Lynch tell the story, it would seem that he is ineluctably closing in on the quarry, the skeleton key to memory. But Bliss and Kandel and Alkon also provide compelling narratives. There is still no direct link showing that LTP indeed turns on the calpain mechanism. And, despite all the encouraging evidence, it is not even universally accepted that LTP has anything to do with memory. It could be no more than a laboratory curiosity. No one has yet figured out a way to teach an animal a fact, then go into its brain and find the neural symbols.

There are many other questions as well. While studies of amnesiacs like R.B. show that the hippocampus is crucial to forming long-term memories, the final storage site for the engrams is believed to be in the area of the brain called the neocortex. Developed fairly late in the course of evolution, this is where the brain's most sophisticated processing is done. Lynch assumes that synaptic mechanisms similar to those uncovered in the hippocampus, where it is so much easier to do experiments, are also used to form engrams in these higher parts of the brain. But the manner in which the hippocampus takes the first fleeting impressions—the short-term memories—left on the neocortex by the senses and solidifies them into long-term memories remains obscure. Larry Squire and Stuart Zola-Morgan suggest that during acquisition the hippocampus acts as a temporary storage site for a "summary sketch," in which different kinds of cortical impressions—the shape of an object, its location, its smell—are somehow bound together into a record of the whole event. Eventually, as the pattern is consolidated and stabilized, it is stored entirely in the neocortex, leaving the hippocampus free to process more impressions.

Slowly the scope of the search is narrowing. Nothing is more paralyzing than infinite possibility. Where there once was a wilderness in which it made sense to strike out in any direction, there are now rough pathways, a vague sense of which rivers to follow, which swamps to avoid—a feel for the lay of the land.

"What you're really seeking are constraints," Lynch explained one day as he prepared to deliver a lecture to an auditorium of undergraduates. "You're seeking things that box you in. That's what separates science from most other human endeavors. Religion is not something where people sit down and say, Well, *if* there were a god then. . . . But science is a constant search for that, for those things that hem you in."

It was a clear summer day, and to avoid distraction he was hiding in a trailer on the campus, a makeshift office that had the virtue of being unmarked and unconnected to the telephone system. Freed from responsibility, he continued to reflect about the nature of this strange science where one uses the brain to understand the brain, working most of the while in a domain that remains invisible.

"What you really need to do the best science is a tremendous tolerance for ambiguity. *You have to be able to tolerate ambiguity.* Because we as creatures are set up for some reason to see cause-and-effect. And what you really wind up doing is tolerating the fact that you have all these assumptions and all these uncertainties, and living with them. And when you really go into a novel area, what do you have to guide you? The more novel it is, the fewer the constraints. For a human being that is a very uncomfortable feeling."

Finally, then, it comes down to believing in one's own creation.

"I often don't believe people appreciate the advantages of the calpain hypothesis," he said. "I don't think people really seize upon the fact that this is the thing that says, This enzyme eating that protein causes memory. And that's the end of our story."

A Brain
in a Box

As the Presynapticists continued to battle the Post-synapticists, and the disciples of calpain mounted their insurgency against the establishment—the worshippers of cyclic AMP—neuroscience seemed reminiscent of the days of the medieval heretics. Was Christ identical in nature with God (*homoousios*, in Greek), as the church insisted, or merely similar (*homoiousios*), as the Donatists believed? Will Durant wrote that these two groups "differed by only an iota in their faith"—homoousios as opposed to homo*i*ousios—but in theology and science, details are everything. If Christ was a divine being then could it be possible that a mortal, Mary, was truly his mother, or, as the Nestorians believed, mother of only his human side? The Monophysites decreed that Christ was purely divine, with no human component. And there were the Eunomians, the Sabellians, the Priscillianists. "We can only mourn over the absurdities for which men have died, and will," Durant lamented.

Not that there was anything absurd about figuring out how neurons store memories. But even in science much of what is taken as truth is rooted in gut instincts that finally must be accepted on faith. In a world where one person's data are another person's noise, the mechanisms put forth to explain memory often seemed like chimeras. Look too closely, and the phenomenon would fly away.

By mid-1990, the details of synaptic plasticity were still up in the air, with the Presynapticists preparing to launch a counterattack. At a meeting at Cold Spring Harbor, New York, two scientists—Richard Tsien of Stanford and Charles Stevens of the Salk Institute—unveiled experimental results which seemed to support the theory that LTP works by turning up neurotransmitter flow in the sending neuron. But the same issue of the journal that reported this development also included a paper that bolstered the Postsynapticists' argument.

Still, on the most fundamental point there was general agreement: Whatever the specifics, there must be *some* way to adjust the strength of synapses, to snap neurons into assemblages that act as symbols for everything we know. But what do these patterns look like? Does it take a hundred neurons to represent the opening bars of a symphony, a thousand, a million? How many bulbs are there in each letter on the scoreboard of memory? What are the symbols and the syntax in the language of the brain?

As the battles over the molecular mechanisms of memory raged on, a few neuroscientists were beginning to wonder if they were becoming too obsessed with the details of brain chemistry. Maybe it was not in these microscopic skirmishes that enlightenment lay. And so they set their sights a little higher. Ultimately, of course, the mechanisms of intelligence do not exist at the level of a single neuron, any more than a computer's powers of computation exist within the electrical eddies inside a single transistor. It was crucial to know how transistors worked, but what about the larger circuitry? With the rough outlines of synaptic change worked out, it began to seem more feasible to study how whole webs of neurons interact to generate human behavior.

In moving up to a higher level of abstraction, neuroscientists were entering a territory where biology merges into psychology, and even linguistics comes into play. The questions they were asking were not about chemistry as much as semiotics, the study of how symbols are assigned to ideas and objects and manipulated in that barely fathomable process we call thinking. How is it that one thing can stand for another? What is the nature of the linguistic transactions our brains carry on with other brains?

Words were once thought of as magic—holy sounds intimately related to the things they evoked. In primitive times, the sign was simply a little picture; it resembled the thing that it meant. Think of pictographs on rocks, or Chinese ideograms. The symbol for man was a little man. In astrology, patterns of stars and planets in the heavens were deemed significant because they looked like things on earth, or like beings in the ancient myths. Mars was red, blood was red, rubies were red; so there must be magical connections between them. Patterns in the sky could somehow affect the worldly things they resembled. The written symbols of language eventually became too abstract to recognize as representations of the real world. In Western alphabets single symbols, letters, came to be used for sounds instead of things. But still, it was long assumed that there was some intimate connection between a man and his name, between the word and the thing it stood for. Viewed this way, texts were not just representational, a string of symbols to be decoded. If words and names were magic, then a scripture or an incantation had power to change things in the world. A man could be cursed, quite literally, by plugging his name into the proper spell.

The world underwent an epistemological revolution when people realized that there is a difference between the signifier and the signified, the word itself—its sound, the way it is written—and what it means. *Dog* doesn't have to mean dog any more than a red light and not a blue light has to mean stop. The connection drawn between the symbol and the thing is arbitrary. A word can be used to stand for anything at all, as long as people agree on the general definition. With the advent of computers, people have become so used to thinking of artificial languages that the arbitrariness of assigning thing to symbol is second nature. There is no reason why 1001001010 should stand for A, B, or Z. Any pattern can be used to represent anything at all.

The arbitrary connection between symbol and thing probably also applies to the languages inside the brain. A configuration of neurons, a hundred, a thousand, a million, all firing at the same time, might stand for *rock*. But there need not be anything special about the pattern. Its exact shape and location probably depend on what neurons happen to be available at the moment it is stamped into place, and what other symbols are already encoded in memory. If the word is learned an hour earlier or an hour later, a very different constellation might be formed.

From one day to another, the pattern that means *rock* may not even consist of the same neurons; it might drift somehow to new locations. Think again of the football scoreboard. B is B no matter which bulbs are used to compose it—and it still is B if it glides across the screen like a letter in the Times Square news ticker. Some researchers suggest that patterns of neurons compete with one another in a Darwinian struggle to see which one gets to represent a thing or an idea in the outside world.

With all these variables coming into play, there is no reason to think that *rock* in my brain looks like *rock* in your brain. It doesn't matter as long as a rock in the world produces some enduring, distinct pattern in your brain, and that upon retrieval it triggers the proper vocal response: "Rock." And my brain responds to those sounds as it would respond to *rock* written in a book or a rock sitting on a table— by lighting up my own distinctive neural pattern. Spoken language then would be a kind of Esperanto, allowing for communication between billions of internal worlds, each with its own mother tongue.

Words are not hard little kernels of meaning. The pattern for *rock*—whatever it happens to be—would also be linked to other patterns, a cluster of neurons that meant *hard* or *rough* or *gray*. It might be linked to patterns that stand for visual images of rocks you particularly remember—one you climbed with ropes and pitons, one that you threw at someone's head. In addition to attributes and images, there must be patterns that represent how a word sounds. For *rock*, this auditory encoding would have to be linked to patterns representing other meanings: what a boat does in oscillating water, a kind of music, a dead actor's first name. In the brain, *rock* would have no definite edges. It would sprawl through the cortex, encroaching into the territories staked out by other words. Once a pattern of neurons was activated it would spread, stimulating patterns that stimulated other patterns. Meaning would ripple through the brain's tissue of associations like the waves from a rock dropped into a pond.

This is all speculation, of course—and speculation based on introspection, which the behaviorists foolishly tried to banish from psychology. Every person's mind is a laboratory. What is needed is a way to test the insights gained from our constant, late-night thought experiments. But how can we find memory patterns in the brain when we can't even watch a single synapse change? There is no way to observe a brain while it is learning, and see the patterns form, to watch a brain while it is remembering, and see one pattern ignite another, retrieving an association.

For want of some kind of cogniscope that would zoom in on a mind at play, some neuroscientists in the mid-1980s decided to try a radically different approach. They decided to make artificial brains and study them. They would put together a system of artificial neurons and synapses and see if they could get it to learn. Then they would open it up, peer inside, and try to determine just what it was that had changed. This, they believed, would provide clues to the workings of real brains.

One way to do this would be to take a few transistors and other components and wire them together so that they acted like a single neuron—signals would enter one end and emerge transformed from the other. Then thousands of these model neurons could be wired together to make a network. In principle this is what the neural modelers did, but instead of making the nets by stringing together physical electronic components, they simulated them on a digital computer. After all, a computer can be programmed to mimic any complex process, a hurricane, a nuclear plant, the economy of Paraguay. Why not the brain? Of course neither the meteorological nor the econometric models work very well. But it is assumed that as we continue to refine them they'll come closer to the mark.

In a model of the brain there would be no need to simulate every detail of neuronal chemistry any more than a weather model has to simulate every water and oxygen molecule in the atmosphere, or an econometric model has to simulate every player in the economy—or, for that matter, every neuron in every player's brain. Modeling is based on the assumption that it is possible to work at a higher level of abstraction where trivial details blur together into more manageable wholes. In an econometric model we can treat each person as a black box, a featureless unit that simply generates and consumes goods and wealth. Actually, economists work at an even more abstract level. A whole country might become a black box, what engineers call an input-output device. Money, energy, and resources flow into one end, goods and services spew out the other.

With all its ionic eddies and enzymes, faithfully simulating a single neuron would take an entire supercomputer. But it shouldn't be necessary to work with so fine a grain. Neuroscientists have assumed for years that when all is said and done, the neuron operates like a fairly simple little computer. It receives signals from other neurons. Some of the signals are excitatory; some are inhibitory. They add and cancel, and if the sum exceeds the neuron's threshold, it fires, sending its own signal to other neurons.

Stripped of its biological complications, a neuron could be considered a black box, just another input-output device. Given a signal, it processes it according to certain rules, and reports, in code, what it has found. It takes input and converts it into output. If neurons are thought of as circles and synapses as lines connecting one to another, a whole network could be laid out piece by piece on a video screen. Numbers could be placed by each line indicating the strengths of the synapses (with positive signs for excitatory connections, minus signs for inhibitory ones). The whole system could be programmed with some version of Hebb's learning rule. There would be no ion channels in the simulation, no sodium and potassium pumps, no receptors, no calpain or cyclic AMP. But in the action of the lines and circles—these simple abstract neurons and synapses—the biochemical machinery would be implicitly accounted for.

Or so the theory goes. For scientists frustrated by the ambiguities of interpreting cellular behavior, what could be more captivating than this notion of toying with simulated neural tissue? Like computer hackers playing with their first artificial universes, these scientists learned the pleasure of absolute control. With the tap of a keyboard they could rewire a network, changing the number of neurons, the density of connections, the firing thresholds. Before long, hundreds of neurobiologists were performing experiments in silica that could never be done in vivo or in vitro. Some were even claiming they had come up with simple networks that worked like human memories— that could not only store information but sort it into categories, that could generalize, drawing on experience to recognize similarities in the world. But the purpose of this new crusade was not to make artificial intelligence. Once the scientists had figured out how to design a neural network that acted like a memory, they could take the blueprints back to their laboratories and look for evidence that similar structures existed in real brains. The computer models would provide clues for what to look for, landmarks for exploring the neural wilderness.

· · ·

One of the first to catch the wave of this avant-garde was Gary Lynch. Toward the end of the 1980s he was spending more and more time outside the laboratory engaged in marathon conversations with a young computer scientist named Richard Granger. Granger had come to Irvine from Yale, where he had studied computer science and psychology with Roger Schank, one of the most prominent theorists in artificial intelligence, the attempt to program computers to think like people.

Most people in A.I. had no interest in neurons, even simulated ones. Instead they would try to write programs that enabled a digital computer to answer questions about a one-paragraph story, for example, or to pick out and name the objects in a cluttered visual scene. The assumption was that there were rules of thought that could be programmed into a machine. To understand the sentences "John hit Alan with a bat" and "A bat bit John," the computer would need rules of syntax and semantics. *John* and *Alan* are names of people. If *hit* appears between two nouns, the first one probably refers to the person or thing that did the hitting. In a sentence about hitting, the word *bat*, especially when preceded by the phrase *with a*, probably refers to a stick, not an animal. But sticks don't bite people. To understand what was what in a scene of building blocks, a vision program would have rules like this: Three lines joining to form an arrow could be the peak of a pyramid or the corner of a block viewed from below.

The lightning-fast, step-by-step precision of a computer bears little resemblance to the workings of a brain. But according to the central dogma of A.I., such distinctions are irrelevant. Intelligence, it was believed, could be translated into rules and implanted into any kind of information processor, as long as it was powerful enough. The nature of the substrate—hardware or wetware—was of no consequence. According to this creed, intelligence was best understood from the top down rather than from the bottom up. It was the software, not the hardware, that counted.

"Almost no one in A.I. knows very much about neurobiology," Granger lamented one day in 1984, not long after he had moved to Southern California. To the A.I. people, the mind was a bunch of programs, and a program can be rigged to run on any kind of computing machine, electronic or biological. "If you asked me a few years ago, when I was in college, I would have said the same thing, but as it turned out I was wrong."

To understand things like memory and perception it helped to step back and look at the brain from a higher vantage point. But Granger had begun to feel that his colleagues in A.I. had moved too far away from the interesting question: how something like neural tissue can learn and remember. He was still more interested in programming computers than in working with microelectrodes. But he saw that network modeling offered a middle way. He wanted his programs to simulate neurons, not the rules by which we sort subjects from objects or triangles from squares. But to make his networks realistic, he was going to have to consort with people who got their fingers wet.

Psychologists and neurobiologists tended to avoid one another, but between neurobiologists and computer scientists there was hardly any communication at all. Each group dwelled in its own hermetically sealed universe. But after several conversations, Granger and Lynch realized that their interests overlapped. They were pursuing the same thing, an understanding of memory. They had simply been coming at it from different directions. By combining their two approaches perhaps they could converge on a theory of memory, one that explained not just how synapses changed but how labyrinths of neurons formed maps of the world.

Their first thought, naturally enough, was to model the hippocampus. Since Lynch began his work on LTP, the hippocampus had become so popular a playing field for memory researchers that whole sessions of the annual Society for Neuroscience conferences were devoted to nothing else. There was a flood of data that could conceivably be incorporated into a model. But despite its nice, layered structure, the hippocampus would be hard to simulate. For one thing, it was deep inside the brain. Speaking in a code that was still inscrutable, it dispatched and received messages to and from other parts of the brain. But it didn't get signals directly from the outside world. By the time it received a signal a great deal of processing had already been done. It wasn't at all clear how you would go about studying and mimicking the hippocampus's behavior. What kind of messages would you send it? What kind of responses would you expect in return?

Instead Granger and Lynch decided to collaborate on a program that roughly mimicked a piece of the brain called the olfactory cortex, where information about smells is processed. Unlike the hippocampus, the olfactory cortex was close to the outside world. It received signals from the olfactory receptors, the sensory neurons in the nasal passage that respond to odors wafting in. There was very little preprocessing to confuse matters. It was fairly clear what a computer model of this piece of cortex should do: take in information about smells and categorize it. Once trained, the network should be able to tell cheese from water and maybe cheddar from provolone. This small section of the brain also offered another advantage. While some parts of the cortex are numbingly complex, the olfactory cortex was fairly simple, especially if the two scientists could get away with ignoring some of the details.

. . .

To a rat, smells are a very important part of the conceptual universe, symbols it uses to represent the lay of the land. Using smells to mark the alley in a maze where a reward was hidden, Lynch and Ursula Staubli had found that they could train rats to recognize a number of complex odors. They made the smells by combining chemicals used by the perfume industry. The rats were quite discriminating. They could learn to recognize a smell made of two or even three primary odors—A, B, and C, for example—and distinguish it from another made of A, B, and D. Somehow the circuitry of the olfactory cortex must represent ABC differently from ABD.

Like all sensory information, smells are carried to the brain in the form of electrical signals. Using electrodes planted in the rats' olfactory circuits, Lynch and Staubli found they could eliminate the real odors altogether and train the animals to tell one pattern of electrical stimulation from another, a signal that represented ABC from a signal that represented ABD. As it turned out, the memories of these "electrical odors" were especially intense if the signals were administered at frequencies equivalent to the theta rhythm, the natural rhythm of LTP.

As Lynch saw it, LTP must be occurring in the olfactory cortex, linking neurons into patterns that stood for smells. The olfactory cortex was a machine, a little computer that processed information that had been converted into a language of electrochemical pulses. If Lynch and Granger could devise an artificial neural net to simulate the olfactory cortex, then perhaps they could get it to act like a simple memory device.

To simplify the problem they decided to simulate a tiny piece of just one of the three layers of the cortex—a hundred of the hundreds of thousands or millions of neurons a rat uses to classify smells. Feeding into the dendrites of this network was a thick bundle of axons that carried signals something like those Lynch believed came from the nose. The neurons in the network were also connected to one another, more or less at random, as seemed to be the case in the olfactory cortex. Some connections were excitatory, some were inhibitory. Some of the cells had outputs that looped back and connected with their inputs, allowing them to stimulate themselves. Within this tangle of connections, synapses could change strength according to rules for LTP similar to those that Lynch had discovered in his experiments with hippocampal slices. By typing commands on a keyboard he and Granger could change the firing thresholds of the neurons, change the density of the interconnections—tinker with different possibilities.

Experiments have shown that in real noses certain cells seem to act as detectors for each of the primary odors. These cells are connected to a structure called the olfactory bulb, which relays the signals to the olfactory cortex through a bundle of axons. In their model, Lynch and Granger arbitrarily assigned five lines in the bundle to each primary odor, A, B, C, or D. The more of these axons that were active, the stronger the smell. A complex odor, like ABD, might look like this: 10011 11001 01100. The first axon is firing, the second two are off, the next two active, and so forth. In other words, odors A and B, represented by the first two groups of digits, are about the same strength; D is a little weaker.

Given an encoding for an odor, the network would begin processing it. Signals would flow from neuron to neuron, stimulating some, inhibiting others. Synapses would change strength, forming new connections. As the odor was presented over and over, a pattern of linked neurons would emerge from the randomness, a structure that represented that particular smell.

To get a better feel for how this would happen, imagine the process in slow motion. And to simplify matters further, focus at first on just one neuron in the network. It receives a burst of impulses from one of the active lines in the axon bundle that brings signals from the nose. The cell's threshold is exceeded, so it fires, sending signals to other neurons. Now these neurons are also receiving signals from other parts of the network—some excitatory, some inhibitory. This causes some of them to fire and send still more signals. Adding another level of complexity is the fact that the connections between the neurons are of different strengths, and the strengths are constantly changing according to the rules put forth by Hebb and elaborated in Larson and Lynch's experiments on LTP.

Finally, this whole chain reaction of neurons firing neurons firing neurons settles down into a stable configuration, with some cells on and some cells off. This pattern of activity is a representation for the odor. The next time the smell is presented, the pattern might be slightly different. But after several trials, an enduring structure is formed, an engram that stands for the smell.

One afternoon, after an all-night session in front of the computer terminal, Lynch described a typical experiment.

"Say you have a hundred neurons in the net, with axons carrying input from the outside world. You say, Okay, now I want to make a stimulus. Let's imagine that these five lines are active, and these five are active, and these two are active. And that's a representation of an odor or something in the environment. And you run it through the net, and you try and see what happens when you train the thing."

After a number of trials, the network learns to recognize the odor, responding to it each time with a characteristic pattern of active neurons, a symbol that could be arbitrarily labeled *Swiss cheese*.

"Then you put in some other odors, and you put in some others, and you start asking questions."

How many odors can the network learn to recognize before it becomes saturated and begins to confuse one with another? How different must two odors be in order for them to be distinguished? It was like having a primitive little brain to play with.

"We can test what the rat does against what the network does, and we can do the same experiments on ourselves. So we can take a known odor made up of four components, and we can say, Okay, suppose I take one of these components out. Do I still smell the same odor? Does a rat smell the same odor? What does the network say? Suppose I have an odor that has four components, and I take another odor that has four components, and I mix them together. What do

I smell, what does the rat smell, what does the network smell?"

One day Lynch and Granger found to their surprise that their network could also do associative learning, grouping similar odors into the same category. Train the network with ABC, and it will learn to respond to the signal with a characteristic pattern of lit-up neurons. Train it with ABD, and it will respond with a different pattern. But once it learns ABE, it reconfigures itself. Since ABC, ABD, and ABE all have something in common, the network groups them together, reacting to each with the same response. Somewhere in its constellation of connections, it has formed a concept.

The eerie thing was that neither Lynch nor Granger had made any effort to build this ability into the device. A crude form of generalization seemed to emerge simply as a consequence of the architecture of the network. And the architecture was based on a structure in the mammalian brain.

As he described it, Lynch could barely contain his excitement. "It sees one cue, and it will go, Oh, that's nice, and it learns it, and it sees another one that has something in common, and it goes, Okay, I learned that. Now it sees a third one that has something in common with the first two, and as though by magic it forms a category—post hoc, see, post hoc! What emerges out of this thing is this category which did not exist before.

"You didn't tell it to form the category—you don't have a format to do that. There's no program that says, Form categories after you've learned three of a kind. It just does it. Yeah, inductively. Isn't that wild?"

More intriguing still, once the network had been trained it could recognize a smell both by its category (cheese) and by its name (Velveeta). Or imagine that ABC represented the smell of a rose. On the first presentation (the first "sniff," Lynch called it), the network would respond with the pattern for the category: flower. On the next sniff, it would be more specific: sweet-smelling flower. On the third sniff, it would get even more specific: rose.

Of course the input and the output of the network could represent anything—colors or sounds instead of smells. Lynch believes the olfactory cortex is a primitive form of circuitry where the ability to store and associate information of all varieties was born. "It's an incredibly ancient kind of wiring," he said. "It's a kind of wiring that goes back to the reptile brain, the first cortex."

Perhaps by grouping smells into categories that aided survival, this primitive neural network gave rise to the first concepts. From there the brain evolved to deal with more abstract categories: red, blue, loud, soft, all the way up to truth and beauty. "Those things are really extraordinary elaborations of this very primitive kind of olfactory processing that was really present before there even was a neocortex," Lynch said.

The olfactory cortex does not work in isolation. In real brains it also sends signals to the hippocampus and other parts of the brain. By lesioning rats, Lynch and Staubli have shown that without the hippocampus, the memories in the olfactory cortex rapidly fade. He speculates that the hippocampus is somehow involved in long-term storage and in associating odors with visual cues to form more complex representations—the smell of an apple combined with its color and shape. But he hopes that those first fleeting, primitive engrams impressed on the olfactory cortex are somehow fundamental, providing a window on how the rest of the complex machinery operates.

"Your hippocampus is only three synapses from the olfactory receptors, from the chemical odorant," he said one day. "Think about it! Three synapses from the air! If we were not all such visual animals, and we were interested in the study of memory—believe me, if you were a dog and you were interested in the study of memory—it would never even occur to you to look at vision."

As the 1990s began, Lynch continued in his role as grand overseer of his labs at the Center for the Neurobiology of Learning and Memory. He was still very much in charge of plotting strategy, designing the experimental regimens he hoped would convince his colleagues that the calpain hypothesis was true. But his mind was moving in new directions. More and more of his time was spent in front of a computer screen. There he would be, hunched over a keyboard, tapping out instructions, setting different firing thresholds, adjusting the LTP curves. Lynch had never been content to keep his sights trained on the lowest level of abstraction. He was always jumping up a level or two, trying to click the details together into a theory. Now here was a way to make theories that were not just talk. Or, for that matter, not just mathematics. Here was a theory of rat thinking that, in a sense, could think.

It was an eerie feeling. There he was, late at night, the sun long gone behind the hills that sheltered the campus from the Pacific, tapping away at a computer terminal and watching the screen, mesmerized.

PART TWO

The Memory Machine

Why, when the singing ended and we turned
Toward the town, tell why the glassy lights,
The lights in the fishing boats at anchor there,
As the night descended, tilting in the air,
Mastered the night and portioned out the sea,
Fixing emblazoned zones and fiery poles,
Arranging, deepening, enchanting night.

—WALLACE STEVENS, "The Idea of Order at Key West"

Theorizing
About Theorizing

AS HE SAT SIPPING WHISKEY in his office high above the campus of Brown University, Leon Cooper began to speculate about whether the world outside his window was really there.

"If you're a complete solipsist, then even the external world is a hypothesis," he said, gazing out at what to all appearances was a spectacular view of the harbor and the surrounding city of Providence, Rhode Island. Over the years the sight had become a familiar presence, so familiar, he said, that he rarely took the time to relish it anymore. Except at night. The light from the sun would fade to blackness, leaving the geometries of artificial light that show the mind of man has been at work imposing orders on a world that does its best to resist them.

Cooper is certainly no solipsist, but during his career as a physicist—and now as a physicist trying to construct a model of human memory—he has often paused to wonder about the elusive interplay between a theory and the world it is meant to describe.

"People act as though we're mining a lode that is already there," he said. "Maybe the world was created by someone with a game plan, and we're discovering the game plan. But to a certain extent we're *inventing* these marvelous orders."

Among the marvelous orders that Leon Cooper has helped to invent is the BCS theory of superconductivity, for which he, Robert Schrieffer, and John Bardeen won the Nobel Prize in physics in 1972. By showing how, at temperatures near absolute zero, electrons can engage in a complex dance and overcome electrical resistance, he and his colleagues explained a phenomenon that had almost seemed to defy the laws of physics. In a supercooled superconducting loop, electricity would essentially flow forever.

Since then Cooper has turned his sights to neuroscience, becoming a leading player in neural network theory, the attempt to use computer simulations to explain how billions of neurons interact to generate the mind. He has also started a company, Nestor Inc., to put his ideas to work. Some of his theories about how societies of neurons learn to make sense of the world are being used by engineers to make software that can identify handwritten characters or spoken sounds. Like most scientists, he assumes that there is some kind of universe out there to theorize about. But how much of the orderliness is out in the world, and how much is in the mind?

"That's one of those deep philosophical questions I'll only discuss drinking whiskey and looking out on the setting sun," he said, repeating a favorite line. "Are there objective 'laws of nature' out there? That's not an easy question."

But it is just the kind of philosophical conundrum that appeals to Leon Cooper, a theorist who spends a great deal of time theorizing about what it means to theorize. Is mathematics invented or discovered? And what about concepts: *tree, dog, floor, neurotransmitter, ion channel, electron, quark, time*? Do they exist in some Platonic phantom zone, waiting to be discovered by that most efficient of pattern recognizers, the human brain? Or does the brain invent the concepts as part of its imperfect attempt to explain how the world works?

In nuclear physics, Cooper's old specialty, this line between the subjective and the objective has become especially fuzzy. As early as high school, students are confronted with the puzzling notion that, according to quantum theory, light is both a particle and a wave. It depends on how you look at it. Stranger still, an electron cannot be said to occupy a certain place at a certain time. Until we cause it to collide with our measuring instrument, it exists in a state of pure potentiality called a probability wave. Even when we measure the electron—when we "collapse the wave function," as the physicists say—we have to settle for knowing either its speed or its position. Not only is it impossible to measure both quantities at the same time; it is meaningless to say that the electron simultaneously has both a position and a velocity. Position or velocity are imposed on it by the experimenter; the observer is an integral part of the experiment. It seems that it was too much to hope that the concepts our brains evolved for surviving on a hostile planet would carry over into the submicroscopic realm of the atom.

This notion that reality is to some extent rooted in the conscious-ness of the observer has spawned a great deal of pseudoscientific nonsense. Bookstore shelves are full of works purporting to use quan-tum theory to explain everything from holistic health to ESP and Zen. Even some physicists have embraced a radical interpretation of quantum theory called the anthropic principle, which holds that the universe's very existence depends on the fact that we are here to observe it.

Most physicists—Cooper among them—consider that absurd. "There are people who make statements that I never could go along with—that somehow the resolution of problems in quantum theory has to do with consciousness. Everything I've written is against that," he said. "To me, that's pure obfuscation—explaining one mystery by introducing another."

One way to expel this quantum mysticism would be to clarify just what we mean by an "observer." And to understand what an observer is, obviously it would help to understand the physical basis of the mind. Wherever one decides to draw the line between the subjective and the objective, it is clear that we don't stand outside creation like gods, with the power to describe the world with perfect objectivity. It is not just that we affect what we observe by the very act of observation; the world is filtered through our senses, and the information is processed and sorted into patterns by our brains. We only see light in a tiny band of the electromagnetic spectrum. We see what our nervous system allows us to see. The universe would be a very different place for a creature whose brain was wired according to another blueprint, who sampled the world through portals we can only imagine.

Science, then, is not a description of the physical world, but a description of how the world interacts with the mind—and how ex-perience is translated into the structures we call memories.

Seen in this light, it makes perfect sense that a Nobelist fresh from Stockholm and an audience with the Swedish king might stray from the straight and narrow of theoretical physics onto a twenty-year trip through the wilds of neuroscience. To understand quantum theory—and, for that matter, physics itself—you have to understand the role of the observer, and to understand the observer you have to understand the mind. That, anyway, is the theory some of Cooper's colleagues put forth to explain his midcareer adjustment. As much as he likes to talk about abstract matters, when it comes to his personal life—his feelings and hopes and ambitions—Leon Cooper is a very private man.

"For some reason, he is very reluctant to talk about that kind of stuff," said Jim Anderson, a psychology professor at Brown and another pioneer of neural network theory. After years of working together on the same small New England campus, in what was until recently a rather obscure field, Anderson still doesn't feel that he knows his colleague very well.

A dapper man in his late fifties who dresses impeccably, Cooper is not one to let down his guard. It is not surprising that so little has been written about him and his groundbreaking work in neuroscience. "I've sat through hours of interviews," he said, "but apparently I just don't get the point." Ask him a question he doesn't feel like answering, and he is apt to wrinkle his nose and say, "Do I have to talk about *that*?"

"He's a little secretive, actually," Anderson said. "He's kind of funny that way. Part of it is that he likes to be a bit of a showman. He likes to work on something and then pop it out at the last minute and say, Look what I've been doing in my spare time. And it's this very well worked out theory."

In recent years, Cooper has taken time from his work on neural networks to theorize about the immune system, which rivals the brain in its ability to recognize and remember significant patterns—in this case, microscopic invaders. He has even come up with a theory regarding AIDS, which he discussed with Robert Gallo, the co-discoverer of the AIDS virus. For a while, he was thinking about whether there is a good theory that would explain cold fusion, should this controversial phenomenon ever be shown to really exist.

In the end, the best way to make sense of Leon Cooper is as a professional theorist, an expert at arranging facts and suppositions into theories, regardless of the domain. Cooper's success at helping to explain superconductivity showed that he had the refined sense of scientific aesthetics that it takes to build a good theory, one that is robust and elegant at the same time.

"A theory is a well-defined structure hopefully in correspondence with what we observe," he said that afternoon in his office. With the conversation turning from the personal to the abstract, his enthusiasm began to grow. "It's an architecture, a cathedral. There is an ancient human longing to impose rational order on a chaotic world. The detective does it, the magician does it. That's why people love Sherlock Holmes. Science came out of magic. Science is the modern expression of what the ancient magician did. The world is a mess, and people want it to be orderly."

To Cooper, a theorist is an artist, someone with a talent for weeding the essential from the inessential and constructing these marvelous orders.

"For a painter, it is his eye. For a musician, it is his ear. Once you separate out the elements, spinning them into a structure is a matter of technique, like counterpoint. To an outsider, it might seem like black magic, but if it's what you do, then it just seems like an everyday thing."

Unlike symphonies, theories of the universe are written in mathematical formulas. But that is simply a convenience, a way to focus the theorist's mind.

"Mathematics is a language," Cooper observed. "As Galileo said, mathematics makes what I say follow from what I said before. Errors are easier to locate, easier to expunge. In a long political argument, someone can slip in some assumptions while you're not looking, or a contradiction may appear unnoticed somewhere along the line. Before you know it, you're very confused. You can reach any conclusion that you want."

Mathematics keeps a thinker honest. Physics, chemistry, and even biology have yielded to mankind's quantitative powers. Cooper sees no reason why all the mushy philosophy and psychology of the past cannot be replaced by a well-structured theory of the mind.

In 1972, just about the time he received his Nobel Prize, he was at work on his first paper on memory. The question has occupied him ever since. He is not interested in memory as a passive act, the simple recording of information. Cooper wonders how the brain perceives patterns, how it stores them, categorizes them, retrieves them on command—how it builds these maps, these theories about the world.

It was an exciting prospect, though one not much appreciated at the time. Here was a world-class physicist with a philosophical bent turning his sights inward. Unraveling the question of memory could lead to no less than a theory of theorists, an explanation of what it is about the matter inside our heads that lets us make sense of the universe.

"I'm really very smart, and I'm somewhat unconventional," he said, finally agreeing to speculate about the geography of his own inner world. "My mind works in its own ways. I know a lot of people who are a hell of a lot smarter, but I'm up there. And I have enormous confidence in the results of my own thinking. That's why I can do things that seem off the mainstream for many years, because I just have absolute self-confidence. I pursue ideas for a long time.

"Now don't misunderstand me—not all of them work. Nobody bats a thousand. But you have to take chances. Otherwise you might as well do something else where you can make much more money. The whole point of this game is the opportunity to do something at a very high intellectual level."

And then, as the day grew longer, he began thinking about where all this research could lead. He is not, after all, just using computers to understand brains. He is figuring out how to build things that can perceive and remember.

"We are on the threshold of constructing machines that will properly be said to think," he said. "And that might be a very frightening prospect. But you have to look at them as machines that enhance our ability to do things we do ourselves. They won't replace people— they'll work with them, and I think that people's attitudes will change from being threatened by these machines to thinking, *How did we ever live without them?*

"We need systems like that for very complex decision-making. You see examples right now of human systems breaking down because they can't handle all of the variables required for very complex situations—many of the environmental problems, many of the problems involving very large amounts of data, where if you had all the time in the world you probably could make rational decisions. But you don't have all the time in the world."

Then, seemingly in obeisance to some inner censor, he paused, not wanting to get carried too far by his enthusiasms.

"If what you are saying is that nobody's going to construct an entity in a cubic foot that does all the things we do, then I'll say, probably that's not going to be done for a long time. It will probably take a long time before anyone constructs a dialysis machine as small as a kidney. You have to clearly distinguish between understanding what's going on and being able to duplicate it.

"But if what you mean is understanding *how* the nervous system learns, how it stores memory, how it organizes itself, how at least in principle the various subsystems interact, what is the origin of consciousness and self-awareness—there you're talking about something else. And those problems are soluble. I think the very last property—consciousness and self-awareness—will be much harder, but it doesn't have to be particularly mysterious. That's something we're all very interested in, but I personally reserve that for late in the evening, when, as Kafka says, you're looking out the window as the sun sets.

"I think we will understand these things and we will understand them quickly. And when we do understand them, suddenly everything will seem simple. Problems that seemed impossible one day will seem trivial the next."

Atomic Choreography

IN 1955, when John Bardeen was looking for a smart young research associate to help him solve the problem of superconductivity, Leon Cooper was working at the Institute for Advanced Study in Princeton, New Jersey. In those days, the institute, which had been the haunt of Einstein, Gödel, Oppenheimer, and Von Neumann, was one of the hottest theoretical think tanks in the world. Chen Ning Yang, who called himself Frank in honor of Benjamin Franklin, would soon be working there with T. D. Lee on a theory that would overthrow what most scientists had been sure was an ironclad law of the universe, the conservation of parity. According to this principle, a particle and its mirror image obey the same laws of physics. By showing that this isn't always true, Yang and Lee helped explain why the universe contains more matter than antimatter, and they won themselves a Nobel Prize.

While some physicists studied the world inside the nucleus, Bardeen worked at a higher level of abstraction. As a materials, or solid-state, physicist, he studied the overwhelming complications of how large systems of atoms behaved. He too was on the verge of a Nobel Prize for his part in the invention of the transistor. Now he was looking for someone to help him solve the mystery of how some metals cooled to temperatures near absolute zero lose all their electrical resistance.

"He had written to Frank Yang," Cooper later recalled, "probably asking if some young fellow at the institute, skilled in the latest and most fashionable theoretical techniques . . . might be diverted from the true religion of high-energy physics, as it was then known, and convinced that it would be of interest to work on a problem of some importance in solid state."

It was the first time Cooper recalls hearing the word *superconductivity*.

After graduating from the Bronx High School of Science, Cooper had been torn between pursuing biology or physics. Physics won, and now, at twenty-five, he had a Ph.D. from Columbia University. For his thesis he had worked on a number of problems in nuclear physics, the most interesting of which involved mu-meson atoms. By taking deuterium—a heavy form of hydrogen—and replacing its single electron with a much heavier particle called a mu-meson, physicists could make a special kind of atom with a radius so small that it could more easily merge with another and undergo a fusion reaction. Some scientists even explored this process, called mu-meson catalysis, as an early version of cold fusion. It was a hard trick to pull off, and the phenomenon has never been of much practical interest for generating energy. But the mu-meson itself was an engaging theoretical problem, as was the work Cooper was doing in quantum field theory. Moving from nuclear to solid-state physics would be the first of the radical shifts that have characterized his career.

"Although John no doubt explained something about the problem, it's unlikely that I absorbed very much," Cooper remembered. "However, since at the time prospects for progress in field theory seemed rather discouraging, I decided to take the plunge."

And so he moved to the University of Illinois at Champaign-Urbana to work with Bardeen and a graduate student, Bob Schrieffer, who had been assigned superconductivity as a thesis problem. On the door of Schrieffer's office was a sign: Institute for Retarded Studies. As Cooper began poring over everything that had been written in the fruitless attempt to explain superconductivity, he began to appreciate how inadequate the problem could make a person feel.

"You have to appreciate that it was regarded not only as difficult but as insoluble," he said. "It had been unsolved for almost fifty years, and I think most of the major physicists had taken a crack at it."

Heisenberg himself had tried and failed. Even the great Richard Feynman found the problem too daunting. In 1956 Cooper learned of a talk that Feynman had recently given to a group of physicists. He had concluded the lecture like this: "And when one works on it—I warn you before you start—one comes up finally to a terrible shock: One discovers that he is too stupid to solve the problem." Coming up with an answer to this riddle would be one of the theoretical coups of the century.

. . .

A typical metal, like copper or aluminum, is built like a crystalline lattice—a vast, submicroscopic framework of ions, atoms that have been stripped of their outer electrons. Loosed from their atomic moorings, the electrons form what physicists call a Fermi sea. This electronic liquid flows through the array of metal ions, conducting electricity.

If the lattice were perfect and still, the metal would have no resistance. The sea of electrons would flow frictionless through the atomic framework. But this state of grace can only be achieved at absolute zero and with absolutely flawless crystals. In the real world, even the slightest amount of heat causes the lattice to vibrate, and the vibrations interfere with the electron flow. How then could it be that in some metals—very cold, to be sure, but not at absolute zero —the electrons managed to overcome this seemingly inevitable resistance and flow unimpeded?

"It took us a while to get the pieces together, and when we started really working on the theory, we worked six months," Cooper said. "I've worked very hard, but I've never worked that hard for that long with such a consistent production of results. Every day saw a new result! It was amazing. It was great just to be there."

Bardeen, Cooper, and Schrieffer believed that under certain circumstances electrons could join forces and move en masse, in just such a way that they eluded resistance. Locked into a complex choreography, an electron could not be scattered by a vibration or an imperfection of the lattice, because the other electrons would conspire to cancel out the disturbance.

But for the theory to work, a fundamental barrier had to be overcome. The universe is believed to be made from two large families of particles: The fermions, which include electrons, protons, and neutrons, are the building blocks of matter. The bosons, on the other hand, are used as carriers for the four forces: electromagnetism, gravity, and the strong and weak nuclear forces. Photons, for example, are bosons that carry electromagnetism. This was the problem: One of the most basic truths about fermions is that they must obey a law of physics called the Pauli exclusion principle. Like all fermions, an electron is identified by a set of quantum numbers that describe its position, velocity, and whether it is spinning clockwise (called an up spin) or counterclockwise (a down spin). According to Pauli's dictum, no two fermions can share the same quantum numbers—they cannot be doing the same thing at the same time. How then could electrons come together and engage in the mass action that led to superconductivity?

To resolve this seeming contradiction, the scientists exploited a loophole in Pauli's law. If two electrons of opposite spin could pair up somehow, together they would act like a boson. And bosons are exempt from Pauli's exclusion principle.

But that presented another problem. Being of like charge, electrons have a natural tendency to push each other away. How could they pair up? Part of the genius of the BCS theory was that it explained how electrons could overcome their mutual repulsion by interacting with the vibrations of the lattice. The very same vibrations that ordinarily caused resistance in a metal would be used to overcome the repulsive force between two electrons. Then, acting in pairs, they could move in the carefully coordinated manner that allowed them to elude resistance. The result was a kind of atomic jujitsu, in which the power of the adversary—the resistance-causing vibrations—was turned against itself. This helped explain a seeming paradox. The worst conductors—lead, for example—made the best superconductors. The vibrations that produced resistance were required to overcome resistance.

. . .

Robert Schrieffer and Leon Cooper were chosen to unveil the new theory in 1957 at a meeting of the American Physical Society in Philadelphia. But Schrieffer was delayed because of a canceled flight, so Cooper presented both the scheduled talks. He was twenty-seven years old.

"It was an instant sensation," he said. "It was really incredible. Everybody was following me around; the room was jammed. I was too young to appreciate how incredible it was. It was dreamlike. But I was still at an age when I thought these things would happen every day."

He still marvels at how quickly it all unfolded. At an age when many people are still wading into their careers, he had guaranteed himself a place in history—as the C in BCS.

"I guess in retrospect they were heroic days, and they felt heroic. I can't say it felt great. It was intensely painful, great and painful at the same time. I suppose the same way you might say running a marathon is painful—even if you come in first.

"We did a piece of work that is probably one of the more important theoretical achievements of the twentieth century," he said. "I'd say that our theory, in terms of its complexity, the kind of problem we solved, what it has led to, ranks just below the major achievements like the general relativity theory and quantum electrodynamics—just below that.

"If you removed any one of us, we could not have produced it. With the three of us it happened, and it happened in a stunning way."

Twenty-seven years later, when he and his wife came across a single-malt whiskey distilled in 1957, he bought a case. "There is no New England winter day so cold," he observed, "that a glass of this elixir does not warm me instantly."

The Structures
of Experience

SUPERCONDUCTIVITY was a hard act to follow.

"I think the only thing that made sense was not to try to follow the act, just to work," Cooper said.

During the next fifteen years, he embellished his ideas on superconductivity and solid-state physics. He wrote papers on the role of the observer in quantum theory. He dabbled in all kinds of things. "I wasn't wedded to solid-state physics," he said. "I really wasn't a solid-state physicist. I tend to wander from one area to another. In the mid-to-late sixties, my early interest in biological things returned— the kind of difficult, deep questions that I've always liked but that you can't really work on day to day."

In 1972, when he, Bardeen, and Schrieffer went to Stockholm to receive the Nobel Prize, the BCS theory was old news to physicists, a revelation to the rest of the world. "There were so many reporters, so many lectures, so many press conferences—limousines pulling up, taking you from one place to another—and you always have things to do. You're being rushed around from lectures, to the king's palace, to the next reception. But I don't remember enjoying it," Cooper said. "I can't even remember it really. I just remember this continual whirl."

Cooper was already preoccupied with his new interest: what it is about the brain that lets us—indeed compels us—to see order in the world. He was beginning to look at physics from the inside out, focusing less on what was going on in the outside world than on how it was all being arranged as memory structures inside the head. In the late 1960s, certain that beginning physics students could benefit from this perspective, he wrote an unusual college textbook: *An Introduction to the Meaning and Structure of Physics*, which is as sprinkled with historical and literary allusions as Cooper's conversations. Starting with Aristotle and working his way, century by century, to the father of the quark, Murray Gell-Mann, Cooper presented physics as mankind's attempt to use mind and memory to organize experience.

Imagine enrolling in a college physics course, expecting the usual problems about dogs watching flowerpots falling from windowsills, or cannons shooting projectiles from trains, and being hit with a passage like this:

"Man comes into the world with a cry: a burst of light, a slap, initiate him into the universe of sensation. Somehow in the mind this raw experience is ordered, and this order is the substance of science. . . . It may be that a newly born child is not aware that the patterns of light, sound, touch, smell, and taste to which he is exposed have their origin in objects outside his mind. He may not know where he ends and something else begins. The first realization that an often-repeated pattern of sensation is another person—mother—is then a discovery whose magnitude is never equaled. Yet, it is a discovery all of us who grow up and function make."

To a college textbook market accustomed to the likes of Sears and Zemansky or Halliday and Resnick—the standard introductions that describe physics as a bunch of equations that might have been found on golden tablets or inscribed on a Coke bottle that fell from the sky—Cooper's book was quite a departure. In its pages, he marveled at what other authors take for granted: that we live in a world orderly enough to be captured by compact theories.

"The gathering of facts without organization would yield a filing cabinet in disarray, a random dictionary, that dull and useless catalogue sometimes confused with science. Yet what is there in experience itself to indicate that order can be found? . . . We have no guarantee—in fact, it is a little surprising—that we can find any relations such as those between the orbit of the moon and the path of a projectile near the surface of the earth. What is there that leads us to believe that an order we might create would be any less complex than the events themselves, that the symbols we write down on paper will somehow permit us not only to know but also to manipulate the world?"

Minds make physics by boiling down complexity into simple, easily described principles. This said as much about the brain as it did about the rest of the universe.

Cooper's book never really caught on. Perhaps it was its misfortune to come out at a time when Renaissance people seeking a deeper view of science were more apt to turn to *Zen and the Art of Motorcycle Maintenance* or *The Tao of Physics.* In Cooper's book the allusions were firmly planted in the cultural mainstream of the West. Or maybe most students preferred their science to come to them packaged and shrink-wrapped as revealed truth. A few professors appreciated Cooper's sophisticated approach. The book is still used in some classrooms almost two decades later, because Cooper started publishing it himself after Harper & Row let it go out of print.

Magnets and Brain Cells

THROUGHOUT THE twentieth century, grand theories of the brain and mind have been put forth by psychologists, philosophers, and an occasional autodidact emerging from years of isolation with the burning conviction that it all has something to do with quantum theory, a new kind of multidimensional mathematics, or a previously undetected energy field. Maybe that is why so many neuroscientists have a sneering distaste for great, overarching theories of the brain, especially coming from people who haven't seen firsthand the overwhelming complexity of neural tissue. Neuroscience has always been theory-shy—it attracts and rewards people with a passion for detail over general principles. As David Hubel, a Nobel Prize–winning neurobiologist, once wrote, neuroanatomists are "a special breed of people, often compulsive and occasionally even semiparanoid." They are obsessed with precision. A few neuroscientists, like John Eccles, one of the grand men of the field, have become so impressed by the complications of the brain that they have been reduced to mysticism. Slowly, one might piece together a theory of how brains worked, Eccles believed, but when it came to the mind, he echoed Descartes: Mind was an incorporeal essence that inhabited the body, the so-called ghost in the machine.

But Cooper had grown up in a much different intellectual milieu. For a theoretical physicist it was second nature to step back from the matter at hand, to see things from a more distant vantage point where the details began to form patterns, like the dots in a newspaper photograph blurring to yield a picture. But sometimes seeing these hidden orders required tools more subtle than common sense. To help solve the riddle of superconductivity, Cooper had used a combination of statistics and quantum theory—what physicists call many-body theory—to describe how billions of electrons could act en masse to overcome electrical resistance. So, he wondered at first, why not use many-body theory to explain how billions of neurons work together to generate the mind?

"One of the things that intrigued me was that you could read in a textbook about the properties of an individual neuron—they were pretty well understood—but when it came to the properties of ensembles of neurons, the properties associated with memory, learning, et cetera, no one had any idea about that. Nobody had any idea about how or where memory was stored. That just seemed unacceptable! And so I had the notion that maybe I could apply my many-body techniques to systems of neurons. That turned out to be an illusion, but it dragged me into the subject.

"And then there was the possibility of being able to think about all kinds of deep problems like the nature of intelligence, memory, learning, consciousness, self-awareness."

When the mind-brain problem was stripped of distracting details, there were some surprising parallels between how billions of atoms interact in a chunk of matter and how billions of neurons interact in the brain. In both cases there is a population of tiny elements that can each be in one of two states. Neurons can be firing or not firing. Atoms can act like magnets with either north pointing up and south pointing down or vice versa. (The polarity depends on the spin of the atom's electrons.)

In a metal, atoms of both polarities are distributed randomly with no discernible pattern; some point up, some point down. But if the atoms all become aligned in the same direction, they reinforce one another and the metal becomes a magnet. This is what physicists call ferromagnetism. If instead the atomic magnets alternate up-down, up-down like in a checkerboard, the material will be an antiferromagnet— an insulator like ceramic, with no magnetic pull.

The possibility that the mass behavior of neurons might be analogous to the magnetic behavior of atoms was suggested as early as the mid-1950s, and the idea has resurfaced periodically ever since. It is not as farfetched as it sounds. Physicists have discovered that ferromagnetism and antiferromagnetism arise because some atoms tend to force their neighbors to align in the same direction, while other atoms tend to cause their neighbors to line up the opposite way. In the brain a neuron that is firing can either stimulate its neighbor (if the synapse between them is excitatory) or suppress it (if the synapse is inhibitory). One neuron inhibiting another would be like an atom causing a neighboring atom to align itself in the opposite direction, in effect canceling the "signal" out. A neuron that fired another neuron would be like two atoms lining up together to double their influence on other atoms.

Imagine billions of tiny magnets suspended in a box filled with lightweight oil. And suppose for the sake of the analogy that, like atoms, some of the magnets interact ferromagnetically and some antiferromagnetically. Shake the box and the magnets will twist and turn, attracting and repelling one another. Finally, they will settle into a state of equilibrium, distributed more or less randomly but with little patches of order. In some regions north-south magnets will alternate with south-north magnets, canceling one another out. In other places magnets will align themselves in parallel, all pointing in the same direction. These regions will act as mega-magnets, influencing other domains. All these millions of local interactions will determine the overall structure inside the box. As they say in solid-state physics, there is short-range order (magnets affecting magnets) and long-range order (domains affecting domains).

Now tap on the outside of the box. Vibrations will ripple through the medium. Tap regularly and maybe some kind of permanent change will occur in the internal structure. A record will be left of the tapping. A memory will be formed.

Obviously, a lot of details would have to be worked out for a theory like this to be very convincing. In this analogy the mass action of neurons is less like the circuitry in a computer than like weather. Using statistics, scientists might study memory and consciousness in much the same manner that a physicist would study temperature, pressure, and the other higher-level phenomena of the atmosphere. The tools of one scientific domain would carry over into another.

. . .

During the next decade, the possible parallels between solid-state physics and neuroscience pulled a few other physicists into the fold. Some scientists began analyzing strange synthetic materials called spin glasses, whose internal structures are as complex as those described in the imaginary box of magnets. Patches of ferromagnetism (atoms that tend to force one another to align in the same direction) are interspersed with patches of antiferromagnetism (where north alternates with south). Could these internal fields of order be something like the structures memory leaves inside the brain?

Cooper himself finally abandoned this notion and struck out in another direction. But his search for parallels between physics and neuroscience convinced him that just because the brain happens to be a piece of biological machinery doesn't mean that the mind is necessarily a biological problem. It could be studied like other complex systems. He was certain that it wasn't necessary to understand all the details of neuroanatomy and neurophysiology to figure out how memory worked. In fact, the people who did understand this stuff might be blinded by all the details. They might just be in need of a professional theorist, someone who could strip away the superfluities and expose the phenomenon's fundamental core.

"You have to know what to look at and what not to look at," Cooper said. "The world is too complicated a place to take it in all at once. Einstein said you should make a problem as simple as possible— but not simpler."

To study Newton's laws of motion, a physicist will make simplifying assumptions, ignoring friction, for example. Once this pure situation is understood, the messiness of the real world can gradually be introduced.

So what would happen, Cooper wondered, if you ignored most of what was going on in a neuron and considered it not as a magnet but as a simple, computer-like device that gathered signals from other neurons, processed them, and sent its own signal out the other end? Using these streamlined units as Tinkertoys, you could snap together simple networks and see if you could get them to do brainlike things.

"To slavishly reproduce a neuron is a waste of time," he said, "because 95 percent of what a neuron does is probably totally unrelated to information processing or storage. It has the whole problem of housekeeping, just staying alive. The problem is not to slavishly reproduce a neuron, the problem is to understand what it is the neuron is doing that has something to do with information. Once you have that, then that is what you try to reproduce."

Some biologists found this notion annoying, to say the least. Someone who has spent years lovingly studying a particular ion channel or receptor is not likely to welcome the idea that the object of his affection could be irrelevant in the greater scheme of things.

"The general reaction of biologists was, Well, who knows, maybe he's a bright physicist, but things like synaptic modification and storing memory over a network in a distributed way—that was considered somewhere between fantasy and total illusion," Cooper said.

Some of his physicist friends were also skeptical. Ever since the heyday of cybernetics after World War II, a small band of researchers had been driven by the idea that they could understand the brain by putting together networks of artificial neurons. While some of these people actually made machines from tubes and transistors, others simulated networks on a digital computer. But by the early 1970s the neural network field had suffered through decades of failure and was in disrepute. Cooper might as well have announced that he was planning to explore the biological underpinnings of the Jungian unconscious or the connections between gravitons and ESP. But he was convinced he could succeed where others had failed. Once he has an idea, it is hard to wrest it away.

"When I first began talking about this, it was treated with a great deal of skepticism from several points of view," he said. "I think that most people, physicists, for example, would have thought that any attempt to understand the brain was interesting but totally premature—you might be able to do that in a thousand years. As a very well-known friend once said to me, 'Good luck.' Besides, it wasn't physics—that's very important to physicists. And people who built machines, who did practical things with computation, thought networks were thoroughly discredited and would never do anything. So it was about as far out as anything I could do.

"So you might ask me why any rational person would do this sort of thing. Well, I suppose I'm not completely rational. It's probably a fault in my character. I have a kind of ferocious individualism, especially in intellectual matters. If I think an idea through and it seems to me that it makes sense, I really don't care what anybody else thinks. That's an arrogant point of view, I suppose. But if the world has come around to your way of thinking several times, you can get away with it."

And so he began wandering alone into an intellectual hinterland.

"You're on the top of one game, and then suddenly you start as a student again. Well, I didn't exactly start as a student, but somehow that didn't bother me that much. The intellectual stimulation, the process of learning itself, is very exciting. I like that part probably better than anything else. Because you feel as though you're moving so rapidly. Sure, you're just rediscovering what everybody else knows, but it's very stimulating intellectually. And if you're in this game, that is one of the things you're in it for.

"There were many things we were not sure about, but to me it seemed so clear, so evident, so obvious that it would be possible to go along this path and get interesting and fruitful results—of course I couldn't tell how long it would take; you can never tell. But I always hoped I would be here to see results. I've never felt it has to be done the next year. I don't work that way. Of course having a position as a tenured professor helps. But still, there is a problem of getting support, and for a while it was difficult. Now it's much easier. The field has become very, very fashionable."

Universal
Machines

As COOPER BEGAN to immerse himself in the literature of neuroscience, he found a rich if checkered history of people who thought the best way to understand brains was to simulate them with machines. One of the first, and probably the most important, was a young British mathematician named Alan Turing. In 1936 he invented a mechanical oracle that came to be called the Turing machine. This imaginary device was deceptively simple in design, its behavior astonishingly complex. It was made up of two simple components: a scanning head that could read and write marks on an endless strip of paper, and a dial like that on the face of a clock. To ask a question, a human would encode it on the tape in a language of Xs and Os and feed it to the machine. It would move the tape left and right, reading marks, printing marks, erasing marks, until it had produced its answer: another string of Xs and Os that the human could now decipher.

The machine could respond to the input because it had been built to obey rules like this: If there is an X in the square you are reading, and the dial is at position 5, then move the tape three squares to the right and reset the dial to position 2. Then the machine would read that mark and follow the proper instruction, continuing this way until it came across a command to stop. Step by step, it would take the input signal and convert it into an output signal.

With the right set of instructions, it was possible to make a machine that could add (given XXX XX, it would say XXXXX) or one that could subtract, or tell if a number was prime (X could mean yes, O could mean no). For complex calculations, part of the tape could be used as a scratch pad, a memory, to temporarily store symbols. In fact, given enough tape and processing time, a machine could be built that would carry out any task that could be defined as a series of precise, step-by-step operations, what mathematicians call algorithms.

But, most interesting of all, Turing realized that the instructions that gave a machine its special character did not have to be wired in; they could also be coded into Xs and Os and fed to the machine along with the data it was to process. There was no need to have different machines for adding and subtracting and multiplying and dividing. Given the proper tape, a single machine could be made to perform any number of functions. Turing called this tabula rasa a universal machine.

Today we use a more mundane label: the digital computer. People use the universal machines Turing imagined half a century ago to process words, sounds, and images as well as numbers—anything that can be translated into a code of Xs and Os, or as we now say, 1s and 0s. With the proper tape—or, rather, a disk containing magnetically written instructions—a personal computer can be programmed to simulate a typewriter, an index card file, a financial spreadsheet, or to play all kinds of video games. For that matter, an IBM PC can be programmed to imitate a Macintosh; given enough memory and time, it can theoretically do anything a Cray supercomputer can. As a computer scientist would put it, all of these machines, running any of these programs, are computationally equivalent; each of them could be imitated by one of Turing's plodding devices, with its scanner and tape, churning through its list of simpleminded rules.

For Turing, it was natural to venture that, given the right instructions, one of his machines could imitate a person so well that it would be impossible to tell the difference. In 1950 he proposed what artificial intelligence researchers now call the Turing test. Put the machine in one room and a person in another, then interrogate the two with a teletype. If you can't tell which is which, Turing proposed, then the machine as well as the person must be said to think. Turing didn't dare suppose that we would decipher the rules of mental behavior anytime soon. And he knew that digital computers would have to become much faster to apply all the rules quickly enough to carry off the masquerade. More interesting was his implication that the brain itself is some kind of Turing machine.

Neuroscientists don't make a habit of reading mathematics journals, so few of them knew much about computer science. But in the years after Turing unveiled his idea, they were finding more and more evidence that the brain is indeed some kind of information processor. If they had known about his theories, they might have been tempted to describe a neuron as a tiny Turing machine: It converts input signals, presumably written in some kind of code, into output signals. On a higher level, a whole brain could be looked at as a Turing machine. The brain receives coded input through its senses, processes it according to the instructions in its programs, and emits output signals that the muscles and glands translate into speech, action— human behavior. The brain also apparently has programs that allow it to respond to input by writing entirely new programs. In other words, it can learn.

Neurophysiology and computer science continued running on parallel tracks until the early 1940s, when a polymath named Warren McCulloch saw a way to bring them together. By trade McCulloch was a neurologist, a physician whose specialty was the nervous system, but he had also been trained in mathematics, philosophy, sym-

bolic logic, and psychology, as well as physiology and medicine—
everything he would need to apply Turing's ideas to the brain. After
a stint in the late 1920s and early 1930s as a doctor at Bellevue in New
York City, he worked at the Rockland State Hospital for the Insane
in Illinois, later moving to the University of Illinois. By the time he
arrived at the Massachusetts Institute of Technology in 1952 to work
in the Research Electronics Laboratory, he had laid the foundations
of neural network theory.

McCulloch also considered himself something of a philosopher.
He liked to call himself an experimental epistemologist. Epistemology
is the philosophy of how the mind gains knowledge about the world.
Few philosophers believed that experiments could tell them much
about this ancient subject. What McCulloch sought, he would later
say, was no less than "a satisfactory explanation of how we know
what we know, stated in terms of the physics and chemistry, the
anatomy and physiology of the biological system." Or, as he also liked
to put it, "What is a number that a man might know it, and what is
man that he might know a number?" He approached this question
by designing thinking machines.

In recent years, McCulloch has become something of a patron
saint among neural network researchers. While he was alive, his eru-
dite essays and lectures (titles include "Why the Mind Is in the Head,"
"Through the Den of the Metaphysician," "Machines That Think and
Want," and, borrowing from Shakespeare, "Where Is Fancy Bred?")
earned him a reputation as one of the last of the Renaissance thinkers:
a neurologist who also happened to know a lot of mathematics and
philosophy, who spent time not only playing with electronic circuits
but treating the mentally ill. Today the most striking and best known
image of the man is a photograph that appears on the cover of his
collection of writings, *Embodiments of Mind*. With his long white beard,
wrinkled brow, and piercing eyes, he looks like a Hollywood version
of a biblical prophet.

. . .

When he began his research, McCulloch was a professor of psychiatry at the University of Illinois, where he collaborated with a young
mathematical prodigy named Walter Pitts. Like McCulloch, Pitts has

become something of a legend in the neural network community. He ran away from home in the late 1930s, when he was fifteen, and came to the University of Chicago to study logic with a passion that astonished his peers. One of his closest friends there was a student named Jerome Lettvin, who would become a well-known neurophysiologist and one of the pioneers in understanding the circuitry of the visual system. Pitts told Lettvin that he had discovered his calling several years earlier in a most unusual way.

"Walter had been chased by a pack of bullies and hid out in a library in Detroit," Lettvin recalled. "He happened to hide in the mathematics section, and he picked up the first volume of Russell and Whitehead's *Principia Mathematica*." In this formidable work, the philosophers Bertrand Russell and Alfred North Whitehead tried to put mathematics on a solid foundation by showing how it could be derived from logic. "For some reason it grabbed him and he spent the next three or four days in the library going through it. He finished the whole thing in a week. In the end he sent a letter to Russell in England pointing out serious questions and problems in the first volume. Russell sent back an invitation to Walter to be his graduate student. He didn't know that at the time he was twelve or thirteen years old."

Several years later Russell came to the University of Chicago. Pitts and Lettvin sat in on his lectures. But Pitts was too much of a recluse to approach the philosopher and identify himself as the one who had sent the letter criticizing Russell's book. "Walter was a curious fellow in that he didn't like to be known at all," Lettvin said. "He really just did not want to have anybody looking at him." Another time, Lettvin said, Pitts annotated a book by the great logician Rudolf Carnap, who was also at Chicago, pointing out some problems. Then he left the copy, unsigned, in Carnap's office.

"Walter had the impression that the explanation of the world lay in logic," Lettvin said. "And logic required suppression of the ego— one had to be completely without self to do it. So he separated from his family; he had an antipathy to signing his name. The point is, Walter was pure mind. He was a gentle and pleasant person whom people automatically liked enormously. But he never talked about himself or his family. Once a year he sent them anonymous Christmas gifts."

In the early 1940s, when Lettvin was a medical student at the

University of Illinois, he met McCulloch. "Warren was a very interesting fellow," Lettvin said. "He looked and sounded like a Restoration nobleman. You can imagine him with a rapier at his side, ready to take on all comers. He liked to write poetry and was given to cryptic conversations. It was a joy to be with him."

Lettvin introduced McCulloch to Pitts. "We all became fast friends," Lettvin said. "At the time Walter was homeless—he lived more or less hand to mouth—and I was anxious to get away from home. So we spent six months living with McCulloch and his wife."

According to legend, McCulloch, an authority on treating disturbed people, was able to put a solid platform under Walter Pitts's feet. During those six months they combined their expertise and showed that it was possible to construct a machine made of artificial electronic neurons that acted something like a brain.

Like all good theorists, McCulloch and Pitts stripped the problem to its essentials. They imagined that a neuron was a simple device that at every tick of the clock added up the inputs it received, both excitatory and inhibitory, and fired if its threshold had been exceeded. Otherwise, it remained silent. There was no reason, they realized, why such a device couldn't be duplicated with a few electrical relays or vacuum tubes. They knew that real neurons were far more complicated. But this simplified model would be the frictionless track with which they could experiment with neural dynamics. Once the principles were understood, the complications could be considered.

In 1943, in a terse act of scientific understatement, they published a paper, "A Logical Calculus of the Ideas Immanent in Nervous Activity," proving that their artificial neurons could be used to make networks that could take any input string—Turing's Xs and Os—and convert it into the proper output. In other words, McCulloch later wrote, "Pitts and I showed that brains were Turing machines and that any Turing machine could be made out of neurons." A neural net, like a computer, could be made to carry out any conceivable algorithm. The programs would not be written on tapes in binary code; they would be implicit in the wiring of the network. The mathematician John Von Neumann later described their discovery like this: "Anything that can be exhaustively and unambiguously described, anything that can be completely and unambiguously put into words, is, ipso facto, realizable by a suitable finite network."

For those who had never taken the idea of the brain as a machine very seriously, the paper was brimming with implications. The input string to a neural net could be a question, the output its answer. Or given a command—"Pick up the book"—a network could respond with signals to the muscles and glands that would result in the appropriate action.

If the input to a network was the thing desired, the output would be a strategy for how to get it. Imagine that a net received a message from a gland saying that the body needed more water. A message from the eyes would say there is water on the table. Or if none was in sight, a message from memory might say water is kept in the refrigerator. Drawing on these sources, a complex neural net—or a number of neural nets working together—would formulate a plan for reducing the difference between the system's current state (thirsty) and the desired state (quenched). A machine that could do this would be engaging in purposeful behavior.

Looked at this way, learning was simply the ability of a machine to change itself, acquiring new behaviors and rules. While a Turing machine might learn by editing the code in its program strings, a neural network could change its internal wiring by adjusting the strength of its synapses.

McCulloch and Pitts didn't immediately suggest all these possibilities; they were implicit in the paper, teased out over the years by them and their followers. There seemed to be little in the way of cognitive behavior that their model could not account for. In a paper published four years later, they showed how to make networks that could identify a musical chord or melody regardless of its pitch, or identify a shape regardless of its size. This was nothing less than the power of abstraction. A minor is A minor whether played in the key of C or G; big circles and little circles are examples of the same concept. In terms of neuroanatomy, the specifics of these particular networks are now believed to be wrong, but the paper demonstrated that a brainlike machine was capable of extracting regularities in the world. Was this how we come to know universals and concepts, they wondered, the so-called Platonic forms?

Finding universals is the whole purpose of science and philosophy. It is, McCulloch would later say, what we mean by "having an idea." If a neural network can note regularities among things—all full moons are round—it could also note regularities among the regularities: All round things have a circumference of 2 pi r.

Most likely, a brain would consist of a whole community of networks, each with different skills. A McCulloch-Pitts network need not traffic only in information that comes in through the senses. It could talk with other networks, examining *their* output for regularities. It could explore its inner world, looking for patterns among its own ideas. It could know what it knows. This, McCulloch wrote, would let the neural machinery have "an idea of ideas, which is what Spinoza calls consciousness, and thus get far away from sensation."

Toward a
Mushier Computer

FED WITH DESCRIPTIONS of computers and brains, the neural nets inside McCulloch's head seemed to abstract similarity after similarity. He could only conclude that these two rare devices were examples of a single universal, the Turing machine. But though he believed that, at their roots, brains and computers were species of the same genus, he was also struck by the surface dissimilarities. Like most of today's computers, the machines that were emerging in the 1940s and 1950s processed information serially. No matter how complex the problem, it was broken into pieces and solved one step at a time by a single central processor. The result was a little like a crowd in a subway station funneling through a single turnstile.

Since the elements of the computer—the microscopic transistors etched on a silicon chip—are now easily capable of switching speeds measured in millionths of a second, serial processing has not been much of a limitation for computers. But neurons have ploddingly slow switching speeds; the time it takes a brain cell to respond to a signal and fire is measured in thousandths of a second. In other words, neurons are a thousand times slower than transistors.

Human reaction times are slower still. The time it takes for a person to recognize a face or the opening bars of a song—or to respond to a traffic light by slamming on the brakes—is measured in tenths of a second. A tenth of a second is only enough time for about a hundred neural firings. If the brain was a serial machine, then its programs could include no more than a few hundred steps, hardly enough for any kind of complex information processing.

The brain seems to make up for the slowness of its components by processing information in parallel. There is no central processor. Millions of neurons can process information at the same time. While a serial computer programmed to process visual images would analyze a scene by sweeping it line by line, translating it into a long string of 1s and os to be sent to a single processor, a parallel machine would take in the image in a single chunk. Using an array of thousands or millions of processors, each one assigned to a single spot of the picture, it would analyze the whole scene simultaneously.

Computers and brains are also different in another fundamental way. Neurons do not march together in the lockstep manner of microchips. In a digital computer, each bit of data—1 or o—must arrive at a transistor at exactly the right microsecond. The brain has no internal clock to ensure such split-second timing. For that matter, a neuron stimulated to its threshold doesn't always fire, it is just more likely to.

Because of all these characteristics, brains are blessed with a quality called graceful degradation. Break a connection inside your personal computer—the one on your desk, not in your head—and it will probably stop working. Lose a few neurons in the cerebral cortex, and the brain seems to work just fine. A certain amount of redundancy seems to be built into the system. Over the years the accumulated death of millions of irreplaceable cells causes the brain to work more slowly, sometimes taking hours where it once took minutes to retrieve a name. But it deteriorates gradually, instead of crashing like a program with a misplaced bit.

It's a mushy kind of information processing that goes on inside the head, no good at all for calculating pi to seventeen decimal places or for instantly retrieving telephone numbers. But for other kinds of tasks—recognizing a face that has aged ten years since the last encounter, reading graffiti, understanding words pronounced in many accents, songs sung in any key—the mushiness is an advantage. Brains are accustomed to dealing with noise, and they have a high tolerance for ambiguity.

. . .

For those who believed that the best way to understand a brain was to build one, the challenge was to make a mushier machine, one that shared the brain's biological strengths and weaknesses. McCulloch and Pitts had laid the foundation with their simple network designs. The question now was whether people could take these principles and use them to weave more complex networks.

Were people smart enough to understand their own brains so well that they could re-create them? It was hard enough to imagine how nature itself had pulled off this mammoth project. As McCulloch liked to point out, the entire library of information in a person's genes is not nearly enough to describe the trillions of connections inside the human neural net. So, he believed, the genetic code must include just rough specifications for the general shape of the brain. Much of the circuitry would start out as a random jumble. After birth, as information began to flow through the senses, the connections would become organized. It was Donald Hebb who suggested in 1949 how the order might arise from the chaos: Learning would change the strength of the synapses, creating circuitry. At the same time, unused connections might die away through a kind of neurological sculpting.

Perhaps people could build brains the same way that nature did, by starting with a child machine and letting it evolve through learning. Experience would be converted into memory structures equivalent to those in the human head.

It is widely believed that the first person to try this approach was Marvin Minsky, who is now known as one of the progenitors of artificial intelligence. In the early 1950s Minsky studied under McCulloch at M.I.T. He took Hebb's metaphor quite literally, soldering together a random network of vacuum tubes that were connected to one another with volume controls like those on a radio. Minsky once described this electronic monstrosity in a *New Yorker* magazine profile by Jeremy Bernstein.

"It had three hundred tubes and a lot of motors," he recalled.

"It needed some automatic electric clutches, which we machined ourselves. The memory of the machine was stored in the positions of its control knobs, 40 of them, and when the machine was learning, it used the clutches to adjust its own knobs. We used a surplus Gyropilot from a B-24 bomber to move the clutches."

The experiment was, at best, inconclusive. With such a jumble of unreliable components, it was never clear just exactly what the machine was doing.

"Because of the random wiring it had a sort of fail-safe characteristic," Minsky said. "If one of the neurons wasn't working, it wouldn't make much difference, and with nearly three hundred tubes and the thousands of connections we had soldered there would usually be something wrong somewhere. . . . I don't think we ever debugged our machine completely, but that didn't matter. By having this crazy random design, it was almost sure to work no matter how you built it."

Completed in 1951, the machine was called SNARC, for Stochastic Neural Analog Reinforcement Calculator ("stochastic" because the synapses controlled how likely it was that a neuron would fire, not the strength of its signal). The network was supposed to simulate rats learning to run a maze.

. . .

For Minsky, SNARC was just one of many college diversions. He quickly abandoned it and moved onto other projects. But for Frank Rosenblatt, who had been one of Minsky's classmates at the Bronx High School of Science, learning machines became an obsession. As a professor at Cornell University in the late 1950s and early 1960s, Rosenblatt tried to make a machine that could learn to read the letters of the alphabet. One way, of course, would be to make up rules for recognizing the characteristic features of each letter. Capital *A*s are pointed or curved on top and have two feet and a crossbar. Then the rules could be programmed into a computer. But Rosenblatt wanted a machine that would make up its own rules. So he invented a device called the perceptron. Starting from a state of randomness, it would evolve into a letter recognizer.

Over the years he tried out a number of different configurations.

But typically a perceptron consisted of three layers. The eye of the machine was a camera lens that projected the image of a letter onto a twenty-by-twenty array of four hundred photocells, each of which would respond with a 1 or a 0 depending on whether it saw light or darkness. The cells in this artificial retina fed signals to the next layer of the machine, a network of 512 electronic neurons much like the ones used by McCulloch and Pitts. Rosenblatt called them accumulators. Each accumulator was randomly connected to as many as forty retinal cells. Some connections were excitatory, pushing the accumulator toward its firing threshold, while others were inhibitory. If an accumulator received enough stimulation from the electronic retina, it would fire, sending its own signal to a third layer consisting of eight output neurons. Connections between these two layers were also random. And they were adjustable, like synapses. Their strengths could be turned up and down using motorized volume controls.

Rosenblatt trained the perceptron by showing it large block letters printed in a standard form. Suppose he began with A. Since the machine started out in a randomly wired state, its initial response was unpredictable. All one could be sure was that one of the eight output neurons would light up. This signal would be arbitrarily designated A. From then on, this was how the machine would be expected to respond to the letter. If it did not, the motors would kick in and the synapses would be readjusted. Next time the perceptron saw A, it would respond more vigorously, sending stronger signals to the appropriate output unit. Then the machine would be trained on another letter, so that it responded by igniting a different output neuron.

If, instead of one, two of the eight output units were used to indicate a letter, it was possible for the machine to learn to respond to each of the twenty-six letters of the alphabet. (For that matter, it would be simple enough to design a machine with twenty-six output neurons instead of eight.) One by one, Rosenblatt would show the machine a set of flashcards. If the perceptron responded correctly then everything was fine; nothing was readjusted. But if the wrong output units responded, he would intervene and tell the machine it had made a mistake. Synapses that had contributed to the incorrect response were automatically turned down; those that should have fired harder were turned up. After seeing fifteen samples of each letter, the machine learned to recognize the whole alphabet.

But where in the machine did the engrams for the letters reside? When it recognized an *A*, any number of accumulators would respond, sending their signals to the response units. This pattern of lit-up neurons was the memory trace for *A*, but it was distributed throughout the network. It was no easy task to analyze the machine's behavior and say what qualities of *A*-ness it had learned to recognize. Each accumulator was, in a sense, a feature detector. But because of the random wiring, it did not recognize obvious characteristics like the angle at the top of an *A* or the bottom loop on a *B*. The "feature" an accumulator recognized—the presence or absence of light at a number of randomly chosen points—had no name. The random configuration also led to a great deal of redundancy. No single accumulator was crucial to a letter's representation. It could be removed from the net-work—a neuron could die—and the system would keep working, a little fuzzier perhaps but still capable of recognizing the letters it had learned. The machine was capable of graceful degradation.

The letter-recognizing perceptron learned by trial and error, re-quiring a human teacher to reward and punish it. Someone had to tell it when it had confused *B* and *A* so it could retune its synapses. In later experiments, Rosenblatt designed a network that learned with-out human intervention. Given random patterns of dots, one after the other, it would classify them into two groups according to regularities it had detected on its own. Rosenblatt lost little time in heralding the virtues of the new device. "For the first time," he proclaimed, "we have a machine which is capable of having original ideas."

The
Exorcists

ROSENBLATT WAS NOTORIOUS for embarrassing his colleagues with ex-travagant claims. But throughout most of the 1960s there was a sense of optimism among the growing number of neural network enthusi-asts, a feeling that these child machines could be trained to become intelligent adults. From a machine that read block letters, one could go to a machine that recognized letters in other typefaces, or even letters that had been written by hand. From a perceptron that could recognize *A*, it seemed an easy step to a perceptron that recognized whole words and then to a perceptron that could read a book. In his writings, Rosenblatt made it clear that he saw no reason why perceptrons could not be trained—or, even better, train themselves—to recognize objects and understand speech. Neural nets, then, would be used not only as models for understanding the brain but for making smarter computers. And that was a guaranteed way of getting money for basic research from corporations and the government.

As it turned out, the perceptron's limitations were much more severe than Rosenblatt imagined. Try as he might, he and the other people working on neural nets could not get the machines to move much beyond their primitive origins. The death blow came when Minsky, who had become disenchanted with the devices, decided to focus his considerable analytical powers on the question of what exactly a perceptron was doing when it learned.

Perceptrons worked well on some problems, badly on others. No one knew why. When a perceptron failed to learn a certain pattern, was that because the task was impossible for that kind of machine? Maybe more layers were necessary, or more neurons, or a higher density of random connections. Or perhaps with more training the network would eventually get the drift, after a trillion or so trials. Rosenblatt proved an important theorem in neural network theory: A perceptron could learn to do anything that it could be programmed to do. In other words, if there was some way to hardwire a neural network to carry out a task—sorting triangles and circles, perhaps—then a randomly wired child network could eventually be trained to do the same thing. But theoretically the teacher could simply run through every possible combination of synaptic adjustments, one by one, until he hit on the configuration that worked. Even for a fairly simple perceptron, the number of possible combinations of adjustments was astronomical. Suppose you had a single neuron with five synapses, each of which could be adjusted on a scale of one to ten. The number of possible states the system could assume would be five to the tenth power. Add a few more neurons, and the system would suffer from what mathematicians call combinatorial explosion. A rat that finally found its way through a maze by systematically trying every possible route could not be said to have learned anything. These machines were clearly capable of some interesting behavior. But Minsky feared that an aura of romantic mysticism had come to surround them, a belief that it was not necessary to understand the details of how intelligence arose from the neural substrate. Just build a random neural net, and let it evolve into a thinking machine. Even if that worked as a strategy for making artificial brains—and Minsky believed that it would not—it told us little about the nature of intelligence.

At a conference in England in 1961, Minsky found someone who shared his skepticism: Seymour Papert, a young South African mathematician who had studied with the Swiss psychologist Jean Piaget. Papert later joined Minsky at M.I.T., where they collaborated on a book that would gain them notoriety as the archenemies of neural networks, the people who, using nothing more than their mathematical skills, managed to kill off a promising field.

Published in 1969, their book, *Perceptrons*, proved beyond doubt that there were a number of important things that Rosenblatt's machine could never do. The purpose of the book, as the authors described it, was "to dispel what we feared to be the first shadows of a 'holistic' or 'gestalt' misconception that would threaten to haunt the fields of engineering and artificial intelligence as it had earlier haunted biology and psychology."

These were fighting words. Accusing a brain researcher of holism was like charging an astronomer with believing in the predictive powers of the zodiac. The whole purpose of perceptron research was to oppose the mystical notion that there is ineffable mental stuff, that the mind is a ghost in the cerebral machinery. But Minsky and Papert made it clear how little regard they had for the opinions of neural net enthusiasts. Most of the writing about perceptrons, they wrote, was "without scientific value." Like self-proclaimed prophets, the two mathematicians were descending from on high to show their benighted colleagues where they had gone wrong.

That, anyway, was how the book was received. The bitterness still lingers. "It had a very dramatic effect," Leon Cooper said twenty years later. "It killed the field." His colleague Jim Anderson agrees: "They kind of torpedoed it."

Over the years, the rhetoric in *Perceptrons* has been analyzed and reanalyzed with the zeal of theologians dissecting a heretical tract. But in spite of its arrogant tone, *Perceptrons* is less a diatribe than a mathematical tour de force.

In the book, Minsky and Papert prove, for example, that a perceptron cannot compute connectivity—that is, it cannot learn to tell which of these figures consists of a single continuous object and which consists of two objects intertwined:

Nor could a perceptron distinguish between rotations of the same figure. It could not tell] from [or —. Nor could it compute parity: Presented with a cluster of objects, it could not tell if the total number was even or odd.

Minsky and Papert's analysis applied only to single-layer machines, those that had only one tier of modifiable synapses. Rosenblatt and other scientists were also experimenting with multilayered networks, assuming that they would be capable of more complex behavior. One layer of neurons would feed through the first tier of adjustable synapses to another layer. Then that layer would feed through a second tier of synapses to the output neurons. But with these so-called hidden layers, it was not clear which synapses contributed to correct answers and which contributed to wrong answers. How would you know which to reward and which to punish? To Minsky and Papert the problem seemed all but insurmountable.

In the end, it was this great leap of faithlessness that most irritated the neural net people. After pages of painstaking analysis about what single-layer perceptrons could and could not do, the authors seemed to be proclaiming, ex cathedra, that these and even worse limitations would hold for many-layered machines. To their critics, this seemed a tip-off that Minsky and Papert were not just proving some interesting theorems, as they liked to claim, but were marshaling their mathematics in support of a very particular point of view. Those who followed their work knew them as champions of an entirely different way of using computers to understand the brain-mind connection.

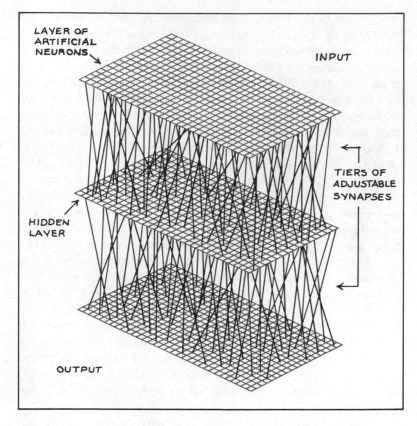

AN ARTIFICIAL NEURAL NETWORK.
Three layers of neurons are randomly connected using two tiers of adjustable synapses.

The name of this alternate approach was artificial intelligence, the attempt to simulate the psychology, not the biology, of the brain.

Systems of Symbols

SINCE THE LATE 1950s the artificial intelligence people had been arguing that trying to simulate intelligence by duplicating the brain's tangled wiring was as silly as trying to make airplanes with flapping feathered wings. It was not *how* the brain worked that was important, they liked to say, but *what* it did. The details of the neurobiology didn't matter; they were just an accident of evolution. Intelligence could be looked at more abstractly, as rules of thought that could be skimmed from their neural substrate and programmed into a digital computer. As Turing had demonstrated, all computers are computationally equivalent. One could be programmed to simulate another. To avoid getting bogged down in all those messy anatomical details, why not concentrate on the software instead of the wetware? Minsky and his followers saw no reason why the mind's programs could not be rigged to run on a digital computer in much the same way that WordPerfect or Lotus 1-2-3 can now be adapted for either a Macintosh or an IBM PC.

No matter how complex the mind, with programs embedded in programs embedded in programs, if it could be broken into a finite number of step-by-step processes, or algorithms, then it should be possible, in principle, to simulate it with a very powerful computer. Like a computer program, everything mental would consist of symbols and algorithms for manipulating them. In this view, it is not just the puzzle-solving, chess-playing kind of intelligence that can be captured with algorithms. Intuition—what we call "having a hunch"—would simply be unconscious processing that we become aware of only after the result is in.

That, of course, begs the question of what consciousness is. But the artificial intelligence people also had an answer for that. Consciousness is the brain's awareness of its own workings. Just as a computer could be programmed with a model of how, say, the bureaucracy of a certain corporation works, it could be given a model of how a dog behaves. With more complex programming, it could conceivably be given a model of the person operating it. So why couldn't it have a model of itself, theories of why it acts the way it does? Then it would not only think but know that it was thinking.

As a professor at M.I.T., Minsky had been encouraging some of the brightest students in the world to concentrate on this new venture. The early results were primitive, but in many ways they seemed more impressive than anything Rosenblatt had done.

Consider Daniel Bobrow's program, Student, which could solve algebra word problems: "If the number of customers Tom gets is twice the square of 20 percent of the number of advertisements he runs, and the number of advertisements he runs is forty-five, what is the number of customers Tom gets?" This is not as trivial as it sounds. To parse the sentence and figure out what meant what, the program had to recognize many patterns: *the number of customers Tom gets* is a unit, the variable x to be solved for; *is* corresponds to the equal sign. To understand other problems the program had to know that people is the plural of person, that a father is a person, and what a perimeter is.

True, the program didn't learn these things. Bobrow provided it with all the rules. In that sense the program was a step backward from the perceptron, but Rosenblatt's machine couldn't begin to learn such detailed, structured knowledge. It was no easy task for a programmer to decide what knowledge was important and, hardest of all, how to represent it in a way that would be useful to the machine.

The implication was that the brain itself consisted of a lot of these kinds of programs. Whether knowledge—Tom, Mary, customers, advertisements, trees, dogs, cars, computers—was represented by patterns of neural firings in a brain or as strings of 1s and 0s in computer registers, intelligence came down to the same thing: weaving symbols into thoughts. What did it matter which kind of loom you used? The symbol not the synapse was the atom of intelligence, Minsky and his followers had come to believe.

Another way to think of the difference is this: A perceptron made up its own sprawling symbols, the engrams, using a logic that often remained obscure to its human inventors. But in an A.I. program, the representations were neatly structured according to theories devised by the programmers. Instead of synaptic strengths and excitatory connections, the A.I. theorists talked about frames and schemata and scripts. All of these were names for symbolic structures in which knowledge was organized in a way that was compact, useful, and very rational. When you walked into a restaurant, you could call on the restaurant script that included everything you knew about restaurants: You exchange food for money; you sit on chairs at tables and use knives, forks, and spoons. Yet the script would be flexible enough to accommodate new situations: a teppanyaki steak house where food is prepared at your table with a flair that is part of the evening's entertainment; a sushi bar where you sit at a counter and are expected to choose living ingredients from a swarming aquarium. One could imagine a tangle of scripts intertwined with scripts. Other researchers concentrated on search strategies: How would you efficiently find the information you wanted without searching every nook and cranny of the memory maze? How do you know that you don't know something without going through your entire mental file?

Looking at the mind from this high vantage point was a liberating feeling. Once the mind's programs had been deciphered, there seemed to be no reason why they couldn't be improved—equipped with vaster memories, sped up using faster machines. The goal was to make intelligence that was superhuman. Freed from the necessity of thinking about the ambiguities of neurophysiology, there were any number of directions in which to go.

"At the same time perceptrons fizzled, we had [James] Slagle's program that got an A in college calculus and Bobrow's program that could do some word problems in high school algebra," Minsky said one day in 1988, reciting some of his students' early accomplishments. "I think a young person would have to be a little bit funny to pursue this other stuff that hadn't worked." Long before the mysteries of the brain were untangled, we could have artificial intelligence, he believed. Maybe it would even take artificial intelligence to decipher the neural wiring; if brains were too stupid to understand brains, maybe we could make machines that could.

While Rosenblatt's perceptrons remained in their infancy, A.I. programs began scoring one tiny triumph after another. Armed with the theorems in Minsky and Papert's book, the A.I. researchers forcefully argued that the situation was unlikely to change. They persuaded the Defense Department, the primary supporter of basic computer science research in the United States, to throw its money behind symbolic programming. Neural nets may have offered a promising path to understanding the brain, but without the immediate lure of practical applications it was impossible to get enough funding to keep the field alive.

In 1972 *Perceptrons* went into its second printing. On the first page of the new edition, added by hand, was a dedication: "In memory of Frank Rosenblatt," who had died alone in a boating accident that some thought was a suicide. But the epitaph might as well have been applied to the entire field. It would be more than a decade before many people started taking neural networks seriously again.

A Model of Memory

"I LIKE TO SAY SOMETIMES that scientific fashion is like fashion in men's and women's clothes," Leon Cooper said. "One year the ties are wide; the next year they're narrow. One year the skirts are high; the next year they're low. And if everyone is wearing a short skirt, you're just hopelessly out of fashion if you're wearing a long skirt. That's the way it sometimes seems with science. You want to be in the middle of what everyone is talking about; you want to be in the mainstream. And the next year it might be something completely different.

"I think people are people whether they're designing clothes or doing science," he said. "One likes to think of science as being a special enterprise or being especially rational. I'm not sure whether or not that's true, but I am sure it's highly exaggerated."

By the time Cooper had won his Nobel and was beginning to think about memory, neural networks were decidedly out of style. Projects like the perceptron were widely seen as quixotic attempts by a benighted few to build a brain one synapse at a time. But Cooper didn't really care. Coming at the problem as a physicist, he had been only vaguely aware of the dispute between artificial intelligence and neural nets. But it seemed clear to him that Minsky and Papert were overstating their case, proving the limits of single-layer perceptrons, then proclaiming that the whole notion of artificial neural nets was intellectually bankrupt.

"It's one of the cardinal errors of intellect," Cooper said. "Although their arguments were very limited, the conclusions they asked you to draw from them were much, much broader."

For those few who stuck with neural net research, many of the arguments in *Perceptrons* were irrelevant. Perceptrons can't compute parity. But neither can we; not at a glance, anyway. With millions of neurons working in parallel we can instantly recognize a human face. But try throwing down a handful of matches and judging whether the total number is even or odd. With some difficulty we somehow force our brain to click into serial mode; like a computer, we count the objects one by one. Minsky and Papert showed that a perceptron cannot compute connectiveness. Again, this makes it more like a real brain. The puzzle on page 144 is as difficult for us as it would be for the perceptron. To tell if there is one continuous object or two, we must trace the lines with a finger. Again, we have to shift from parallel to serial mode. It is the parallel tasks, like recognizing faces, that seem natural to us. It's astounding the difficulty that otherwise intelligent people can have dividing up a restaurant check. The brain's ability to struggle through problems, step by step, is a skill that must have come fairly late in evolutionary life.

In an ideal world where science was purely rational, both the Minskys and the Rosenblatts would have coexisted at roughly equal strength, studying these two important kinds of intelligence. The neural net researchers would study our more primitive but powerful abilities at parallel processing, while the A.I. people would learn how brains were able with great difficulty to simulate serial machines, dividing two numbers with the help of paper and pen. Proceeding from both the top down and the bottom up, scientists might eventually converge on a theory of the brain and mind.

Instead the artificial intelligence people continued to dismiss the neural net people as holists, as though they were trying to conjure intelligence out of mysterious synergistic interactions. The neural net people accused the artificial intelligence people of the cardinal sin of dualism, treating such abstractions as symbols and rules and scripts and schemata as though they were things in themselves, "mind stuff" that could exist without a brain.

"You had something that became an ideological argument," Cooper said. "People were screaming at each other. But it was an ideological argument with its roots in economics, because people were fighting about funding. I just never saw any intellectual conflict. But people tend to take sides, especially when you're competing for limited resources."

It would be years before the two sides realized that the distinction between their approaches was mostly a matter of emphasis, that they were chipping away at different levels of the same problem.

. . .

For Cooper, the choice of directions was obvious. Fashionable or not, he didn't much care how the algorithms of the mind could be abstracted and implanted in any kind of information processor. He wanted to know how real brains worked, and how they changed with experience.

In the early 1970s Cooper began discussing the problem with one of his graduate students at Brown, an aspiring physicist named Menasche Nass. They had read a paper by the psychologist H. C. Longuet-Higgins comparing memory to holograms. The fact that these laser-generated photographs could be cut into halves, quarters, eighths, or sixteenths, with each piece retaining the whole image, seemed like the kind of graceful degradation exhibited by the brain. But holograms weren't very convincing biologically. "I asked Nass to try to cook up a more physiologically acceptable model over the summer," Cooper said. "And as graduate students will often do, he came in at the end of the summer and said, 'I found a model in the literature.' "

What Nass had unearthed was a recent paper by James Anderson, a young neurophysiologist who was among the faithful few still studying neural networks. In 1969 and 1970 he had published papers showing how a network of artificial neurons might operate like a memory, one that shared strengths and weaknesses with a human brain. Nass was so excited by the paper that he went to New York, where Anderson was a postdoc at Rockefeller University, and asked for more information.

"He came into my office one day," Anderson recalled, "and said, 'Tell me everything you know about memory.' " He gave him some papers, "along with a lot of half-baked speculation," to take back to Brown University. Shortly afterward, Cooper and Anderson got together for the first of what would become a series of long conversations. Anderson remembers attending a conference with Cooper in New England in 1972. "Some of our more exciting discussions were conducted while driving very rapidly over the twisty roads of New Hampshire in Cooper's gray Camaro."

Anderson could hardly have been more different from Leon Cooper. A bearded man who prefers flannel shirts to suits, he was as outspoken and irreverent as Cooper was reserved. He had studied neurophysiology at M.I.T. during the 1960s, when people were still enthusiastic about perceptrons. "There was a lot of excitement then," he recalled one afternoon, sitting in front of a computer terminal in his office at Brown. "People really felt they were getting at some of the ideas that were used computationally by the brain."

By the early 1950s, McCulloch and Pitts, along with their friend Jerome Lettvin, had migrated to M.I.T., joining a faculty that included people like Claude Shannon, the inventor of information theory, and Norbert Wiener, the flamboyant founder of cybernetics. During the next two decades these scientists probably did more than anyone to firmly establish the computer metaphor of the brain. Anderson was to become part of the second generation, a group of young scientists for whom the parallels between brains and machines were second nature.

"I actually saw Warren McCulloch a few times when I was in graduate school," Anderson recalled. "He was really a funny guy. Have you ever seen his picture on the cover of that book, with the long white beard? He was almost a god there, like in one of those Michelangelo pictures. And I think that was really the look that he cultivated. The only thing that spoiled it, actually, was that he was a very heavy smoker and there was a yellow streak down the beard. Except for that, he was very impressive.

"I also saw Walter Pitts, who was a very furtive figure. He had a very sad end. He died an alcoholic in a Cambridge rooming house,

I heard. McCulloch kind of stabilized him, I think, and got him to do the good stuff he did. I was told by someone that McCulloch had a habit of picking up brilliant, damaged people and stabilizing them. And he was able to get them to do remarkable things. I think Pitts was like that."

In fact, shortly after they came to M.I.T., Pitts and McCulloch had a falling out with Wiener that was never resolved. Wiener, who was famous for his mood swings, had become something of a father figure to Pitts. As Lettvin remembers, losing Wiener "came as a deadly blow to Walter. He had put up Wiener as the father he never had. He never recovered. From then on, it was downhill. There was nothing any of us could do to pull him back."

In the late 1950s, Pitts burned all his unpublished papers on logic and mathematics. He drank heavily and used his skills as a chemist to make exotic compounds, which he tried out on himself. "A group of us would go hunting for him every night to be sure he was still alive," Lettvin remembers. "We would go searching the taverns."

Anderson recalls seeing Pitts toward the end of his life at a course taught by Lettvin. "It was the next-to-last class of the semester," Anderson said. "And the way I remember it was that Pitts sat up in the front, and Lettvin was talking about how hard it was to figure out the nervous system, what an appallingly difficult job it was. And he'd say, 'Isn't that right, Walter?' And Walter would say, 'Yes, Jerry, that's right.' And Lettvin would say some more depressing things—how it was impossible to figure it out this millennium—and say, 'Isn't that right, Walter?' And Walter would say, 'Yes, Jerry, that's right.' And then at the end of the class Lettvin said, 'Okay it's hopeless. Let's cancel the last class.' And that was the last class of the term."

Pitts died in the late 1960s. Two brothers came to claim the body. It was the first time Lettvin knew that they existed.

. . .

After earning a Ph.D. in neurophysiology with a thesis on Aplysia, Anderson moved on to UCLA for postdoctorate work in the late 1960s. "Although work on Aplysia was interesting," he observed, "and they do have colorful insides, there seemed no immediate possibility of using a very stupid animal like Aplysia to figure out what

I thought were the important questions: How does the human brain work, and what does it do?"

And so he began turning his attention to neural networks, poring through the largely unread body of literature that had accumulated since the early days of McCulloch and Pitts.

"I always thought of myself in the position of the audience," he said. "There were all these people out there generating data, and somebody actually has to sit down and figure out what it all means. This was a difficult thing to do. You had a bad chance of being totally wrong, in which case your chances of getting a job were practically zero. So it was a very high-risk occupation. But back in the late sixties nobody cared about that. You figured you just sort of do what you want. People were really impractical in a lot of respects, but that was okay. It was fun to sit around and talk about this stuff. Because it was very unpopular at the time. That was about the time that Minsky and Papert's book came out and all these neural network models were kind of in decline anyway."

Swept up in the cultural experimentation of the 1960s, Anderson cultivated an image as an iconoclast. While Cooper enjoyed allusions to classical philosophy, Anderson developed a penchant for prefacing his papers with epigraphs from the wisdom of the East:

"All that we are is the result of what we have thought: it is founded on our thoughts, it is made up of our thoughts"—DHAMMAPADA, I. I.

"In the beginner's mind there are many possibilities, but in the expert's there are few"—SHUNRYU SUZUKI, 1970.

. . .

By the time he met Cooper, Anderson had developed a memory model that he called the linear associator. It was an early version of this work that Cooper's graduate student had run across. As inspiring as he found the early neural net research, Anderson was bothered by the fact that Rosenblatt and McCulloch used neurons that were a lot more like transistors or relays than real brain cells. Of course the scientists knew they were oversimplying, striving to cut through to the essence, but Anderson felt they had made things a bit too simple. Their neurons were all-or-nothing devices. They summed up the incoming signals, then fired if their thresholds had been crossed. Otherwise, they remained silent.

But real neurons seem to be what engineers call analog devices; they have a whole range of possible outputs. When it isn't receiving input, a neuron will slowly fire, recharge, fire, giving rise to what is called its resting rhythm. But turn up the volume of the incoming signals, and the rhythm increases. In other words, neurons seem to signal intensity with frequency. A physiologist named Vernon Mountcastle had shown that if he pressed areas on a monkey's palm, certain brain cells would fire faster and faster, depending on the amount of pressure. The brightness of light is also believed to be signaled this way: the stronger the input, the faster the firing rate.

Taking these factors into account, Anderson designed a network with each neuron in the first layer randomly connected to many neurons in the second layer through Hebb synapses. There are a number of places in the brain where one group of neurons projects onto another like this. Anderson showed that this kind of structure could do some rather sophisticated information processing. Like his predecessors, he assumed that the brain represented the phenomena of the world—sounds, smells, arrangements of light—by patterns of firing neurons. In the older digital nets, these patterns looked like this: 1010101010001. Neurons were either on or off. In Anderson's analog net, there were more possible values; 3873846597 would represent a group of ten neurons, each firing at different frequencies. The signal, called a vector, might represent the intensity of light registered by each cell in a tiny patch of the retina or the amount of pressure on a cluster of receptors in the hand. For that matter, it could be an arbitrary internal code that stood for most anything.

Anderson found that if he presented two vectors to his simple network, sending one string of numbers to the first layer and the other string to the second layer, he could train it to associate them. Following Hebb's rule, the synapses would change strength, forging new circuitry. Afterward, if one vector was given to the network it would respond by spitting out the other. Suppose one vector meant Prague and the other meant Czechoslovakia. The machine would learn the association.

Then he would train the network with another pair of vectors and another. Each time, the network would readjust the strengths of its synapses, learning the new association without wiping out the previous ones. What started as a machine that could associate one pair of signals could now associate two, then three, then four. One could imagine giant versions of the network that would process vectors thousands or millions of digits long. One vector might represent a person's face, the second the sound of his name. One vector would recall the other almost instantaneously, giving the machine an enormous advantage over a conventional digital computer, with its single central processor. With Anderson's net there was no need to search through an entire memory bank, item by item, looking for information that might not even be there. The neural net automatically responded to familiar vectors, ignored unfamiliar ones.

The network was also capable of recognition. A vector didn't have to be associated with another vector. By training the two layers with the same vector, a pattern could be associated with itself. In the future, the system would know it had seen the pattern before. Again, there would be no need for a systematic memory search.

At some point the network would become saturated with associations. With too much learning, confusion set in. Similar vectors would blur together, a phenomenon that engineers call cross-talk. Mary might get mixed up with Jerry or Larry—the same kind of mistake that a human brain might make. It seemed that a network like this could store about 10 or 20 percent as many associations as it had neurons. Anyone who has lived past thirty knows how old memories get mashed together. Could this be a function of saturation and neural death?

The network also seemed to share one of the brain's great strengths, a tolerance for ambiguity. It could recognize a vector it had learned even if part of it was missing or distorted. Once it had learned 3572838, it might respond the same way to 3572833 or 35728. We recognize faces wearing a variety of expressions.

Anderson also speculated about how his linear associator could recognize patterns in time as well as in space—melodies or the sequence of tiny bumps sensed by a finger as it runs over a line of Braille. Here he borrowed an idea from McCulloch called time-delay lines. Imagine a five-note melody coming into a network from some other area of the brain. To analyze the whole pattern there would have to be some means of saving the early notes while waiting for the later ones to arrive. One way to do this would be to channel each note through a different input line, each with a different velocity of transmission. The first note would be delayed the longest, the second note delayed a bit less, the third note even less. The final note would hardly be delayed at all. As a result, all the notes would arrive at the input layer at the same time. The melody could be turned into a vector for processing.

In all these cases, it was difficult to point to just where in the network a memory was stored. Each association consisted of a pattern of firing neurons that was spread across the network, overlapping with other patterns. But these constellations were apparent only when the memory was activated: when the network, given the vector for the face of Sam or Nancy, responded with the sound of the name. Until then the engram lay dormant, implicit in the connection strengths of the synapses.

Cooper was impressed enough by Anderson's work to invite him to Brown. "People in this area always have funny institutional backgrounds," Anderson said, "so I ended up at first being part of the applied math department. And I wasn't really a mathematician, so I wasn't too happy. Then I ended up in psychology, which was much nicer." When Cooper formed the Center for Neural Science, one of the first interdisciplinary departments for the study of brain and mind, Anderson became part of that. He has been at Brown ever since.

Animal Logic

IN THE BEGINNING, Cooper had assumed that studying the mass interactions of brain cells would require abstruse mathematics like many-body theory. The linear associator had the virtue of being neurologically realistic yet easy to analyze. It could be understood with nothing more complicated than algebra. The theory was still oversimplified. Neurons aren't as linear as Anderson decided to pretend, smoothly changing their output frequency in response to the intensity of the input. But it seemed like a good approximation. Like a carpenter with a new set of tools, Cooper began, almost automatically, to construct a theory. What would it mean, he wondered, if people had the rough equivalent of linear associators inside their heads?

With neural networks, he realized, what seems a problem of biology—how the brain gives rise to thought—is reduced to a problem of computation. The language of the brain could be thought of as vectors, rows of numbers like 5543722 representing the firing rates of a cluster of neurons. The other important parameter, the strengths of each of the synapses in a neural network, could be represented by an array of numbers called a matrix. A simple neural network might look like this:

24234234
23432423
41221234
12433432

Viewed this way, a network can be easily analyzed. The rule to keep in mind is this: A vector multiplied by a matrix yields another vector. This is just another way of saying that a neural net takes one string of symbols (remember Turing's Xs and Os) and converts it into another string. A matrix then can be thought of as a device that maps one vector onto another. In Anderson's model the network started with its synaptic weights—the numbers in the rows and columns of the matrix—set at random. Each time he trained the network with a pair of vectors, the values in the array would change. What finally resulted was a matrix that acted as a multiplier in a number of different equations. Each input vector multiplied by the matrix yielded the appropriate output vector.

Vectors can also be graphed as lines. This is easiest to visualize in the trivial case of a vector with two components, say 7 and 6. If the first component is on the x-axis and the second on the y-axis, we get a line that looks like the graph opposite.

A three-component vector would require a three-dimensional graph. Larger vectors can only be mapped in multiple dimensions— what mathematicians call hyperspace. But for all practical purposes they can be thought of as lines. No matter how high and unfathomable the dimension, mathematicians speak of vectors that point in the same direction as "parallel," those that point at right angles as "orthogonal."

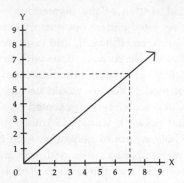

A TWO-DIMENSIONAL VECTOR REPRESENTING 7,6.

Cooper assumed that the brain would use similar vectors to code for similar things. The vector for *white cat* would be more or less parallel to the vector for *gray cat*. On the other hand, the vector for *tricycle* would be orthogonal to the cat vectors.

In general, the smaller the angle between two vectors (the more parallel they are), the more likely it is that a network will confuse them. The larger the angle (the more orthogonal they are), the easier it is to tell them apart. Just how discriminating a network is depends on several factors: the number of neurons, the density of connections, the range of possible firing rates. Obviously, it would be helpful to have some neural networks tuned to lump white cats with gray cats, or Camaros with Mustangs, and others that could tell them apart. We need to know to avoid all cars when crossing the street, but we don't want to get into the wrong one at the parking lot. In some people's brains, Bach and Beethoven might be parallel—they don't listen to classical music much and don't make the distinction. In other people's brains, the two vectors would be most orthogonal. A biologist would never confuse the vectors for amino acid and nucleic acid, but a journalist under pressure might type one term when he meant the other.

• • •

In 1973 Cooper described some of his ideas in a paper called "A Possible Organization of Animal Memory and Learning." In his physics textbook, he had written of the mysterious way that an infant learns to recognize repeated patterns of sensory experience—the combination of light, sound, smell, touch, and taste that he or she comes to associate with Mother. His conversations with Anderson had begun to convince him that this process of induction could be explained with the linear associator model. Vectors would become linked to vectors, forming a complex that somehow represented *Mother*. It wasn't clear how a single neural network would do this. Probably the process would involve a whole swarm of networks, each tuned to recognize different features, all comparing notes with one another.

Cooper also believed that the tendency of a network to suffer from cross-talk and confuse similar, parallel vectors—to respond to 3457677 and 3457777 in the same manner—could be the basis of the ability to generalize. Other women could be recognized as mothers even though they were not exactly like one's own. A digital computer that suffered from cross-talk would be a bad machine. We don't want the IRS file for Social Security number 585-64-1361 to be confused with 585-64-1331. But a network that lumped together cats of different colors could be said to generalize. Philosophers might debate in what cosmic realm the Platonic forms lie. But as far as Cooper was concerned, a network that had learned about gray cats, then responded to white cats and black cats in the same way, could be said to have acquired the concept *cat*. Or suppose the vector for gray cat was associated with the vector for *meow*. Because of cross-talk the network would also know that white cats and striped cats say meow. The sin of jumping to conclusions and the power of logical induction seemed one and the same.

"If the world is properly ordered, an animal system which 'jumps to conclusions' . . . may be better able to adapt and react to the hazards of its environment and thus survive," Cooper wrote. "The animal philosopher sophisticated enough to argue 'The tiger ate my friend but that does not allow me to conclude that he might want to eat me' might then be a recent development whose survival depends on other less sophisticated animals who jump to conclusions." Most creatures will not hesitate to run roughshod over the laws of logic and, with the most meager evidence, stereotype all tigers as potential killers. When the evolutionary pressures abated, people were able to develop more refined skills like logical deduction. But induction itself seemed to be rooted in the very architecture of the brain.

This was a far cry from the work on inductive learning that was being done in artificial intelligence. At about the same time that Cooper was writing his paper, Patrick Winston at M.I.T. was working on a program that could learn simple concepts like *arch*. To teach his program, Winston would present it with various arrangements of building blocks. Two long blocks standing upright with a third block lying on top was an arch, he would tell the machine. But if the supports were touching—a T formation—then it was almost but not quite an arch. A group of three blocks lying on their sides was not an arch at all. Following rules Winston had provided, the program would compare arches with nonarches and near-misses, searching for similarities and differences. Finally, it would learn a rule: An arch consisted of two supports that were not touching and another block lying on top.

In Winston's program, all the knowledge—that which was learned and that which was provided to the system by the programmer—was explicit and orderly. The program was tailored to a single problem. Programs like Winston's helped demystify learning by showing how induction could be done with algorithms. But the networks Cooper was exploring didn't have to be given specific learning rules. By associating vectors with vectors and classes of vectors with classes of vectors, neural networks could be used to weave together "a fabric of events and connections," Cooper surmised. They would create a map of reality. He described the possibility like this: "The system is placed in an environment and, without any search procedures, forms an internal representation of its external world."

Resonance
and Reality

THROUGHOUT THE 1970s, Cooper, Anderson, and the handful of people who hadn't been scared away from neural networks continued to publish papers and gather for small conferences. Though their forces were small, they had at least convinced themselves that they were glimpsing fundamental truths about brains. "People always feel that this was the Dark Ages," Anderson said, "and that neural net people sort of slunk around and hid their faces from the light. But it really wasn't like that at all."

Cooper and Anderson soon discovered that they were not the only people working with linear associators. The same year Anderson unveiled his network, a Finnish computer scientist named Teuvo Kohonen had published the same model in the *IEEE Transactions on Computers*. Few neuroscientists read publications of the Institute of Electrical and Electronics Engineers, so it would be several years before Kohonen and Anderson learned of each other's existence. The fact that the same model for memory had crystallized in two such different mediums, neuroscience and computer science, made the idea seem that much more robust. Neural networks were generating interest not only among people trying to understand the brain but also among those trying to build better computers.

The networks were still vastly oversimplified. But during the next few years scientists tried to close the gap between the real and the abstract. Two physicists, W. A. Little and Gordon Shaw, experimented with another degree of realism and mushiness when they designed a network with probabilistic neurons. The neurons were digital, not analog; they either fired or didn't fire. But as in the brain, a neuron wouldn't always fire when its threshold was crossed; it would just be more likely to. At the same time, a neuron that was below threshold had a slight chance of firing anyway.

As scientists experimented with more and more complex networks, the mathematics naturally grew more abstruse. Some of the most mathematically sophisticated models were made by Stephen Grossberg, who worked with nonlinear neurons. In a linear neuron, like those in Anderson's and Cooper's models, output frequency varies smoothly according to the strength of input; if one is graphed against the other, the result is a straight line. With real neurons, the relationship is quirkier and harder to analyze. In Grossberg's networks, a neuron's input-output graph was S-shaped. The networks also included feedback loops, which made them especially difficult to analyze. In fact, as Anderson once noted, few people in the field understood exactly what Grossberg was trying to do.

From neuroscience and computer science, this new way of thinking about memory moved outward, establishing niches in other fields. Interest was especially strong among people called mathematical psychologists, who tried to come up with systems of equations that explained mental behavior. "They were very open to these novel ideas," Anderson said. "They knew how hard the problem was. They had good data, but they did not have models that were very good at explaining it. So there was a very ready audience."

Anderson became one of the first neuroscientists to work closely with psychologists. At first the exchanges were difficult as two foreign cultures grappled for a common language. But Anderson found that he could produce the kind of results that impressed psychologists. He would take one of those familiar old psych lab experiments and run it with a neural net as the guinea pig instead of a college freshman or a rat. One of his first projects was to design a network that could be trained to memorize and recite the items on a list, making the same kinds of mistakes that people do. He also worked on a network that could recognize different phonemes, a small step toward the distant goal of using machines to understand speech.

While a few researchers actually built neural nets out of transistors and wires, most simulated them on digital computers. Every new breakthrough in neural nets was a testament to Turing's notion that at the highest level of abstraction all information processors, natural or artificial, are equivalent. Each new model stood as further evidence that the brain is some sort of computer. But the network researchers were going a step further than the people in A.I. by showing what the actual wiring might be like.

Inspired by artificial intelligence, many psychologists were seized by the idea that the churnings of the mind could be thought of as algorithms, step-by-step procedures that could be embodied in computer programs. But when eyed too closely, the metaphor became strained. In a computer, memory and processing are completely separate functions—different boxes on the architectural plans. Every parcel of information is assigned an address and stored in an array of memory chips or on a magnetic disk. When the central processor needs the information, it must be summoned from its numbered cell. The computer has to know where the memory is stored in order to retrieve it.

Sure, you could use a serial machine to simulate psychological behavior without paying attention to anatomy and physiology. But for those who wanted to know what *kind* of computer we have in our heads, the programs of the neural net people, which contained simulated neurons and synapses instead of schemata, scripts, and if-then rules, seemed much closer to the mark.

"Many ways to store and retrieve information exist: filing cabinets, libraries and computers. But the fact that an animal's memory is held in a living structure and is successfully utilized, even though the animal may have no idea where his memories are stored or how they are ordered, places special requirements on theory," Cooper wrote. "A basic problem in understanding the organization of memory in a biological system is to understand how a vast quantity of information can be stored and recalled by a system composed of vulnerable and relatively unreliable elements, with no knowledge of where the information has been filed."

By showing how memory and processing could all take place within the same system, Cooper refined the computer metaphor in a way that some psychologists found appealing. Throughout the computer revolution of the 1950s and 1960s one could find psychology books with diagrams of brains that looked like flowcharts. Laying out the mind in such neat parcels helped combat holism, the notion that there was something mysterious and indescribable about thought. But to many scientists the diagrams had too schematic a flavor. They just didn't feel mindlike.

As Cooper had shown in his paper on animal logic, there was no need for a memory box that stored data to be processed, a logic box that drew inferences, or a generalization box programmed with rules for inductive reasoning. Memory and logic were intertwined, woven so tightly together that it was all but impossible to tease them apart.

Nor was there a need to draw a sharp line separating perception from memory. On the psychologists' maps of the mind, these two artificially delineated regions could also be merged into one.

Looking out the window at the ocean, we might notice a bright light in the night sky hovering on the horizon. Deep inside the brain one neural network responds to this vector, dismissing it as just another star. But its intense brightness causes another network to guess that it is Venus. Then the light starts getting bigger, brighter, creating a different vector, a different set of firing patterns. Another network associates this configuration with approaching headlights on a freeway. Then two more lights appear, green and red. Networks that interpret these colors feed into other networks; the pattern for *stop light* weakly responds. All over the brain, networks are talking to networks, entertaining competing hypotheses. Then comes the roar, and suddenly we know what it is. The noise vector, the growing-white-light vector, the red-and-green-light vector all converge on the network—or network of networks—that says *airplane*.

Memory creates a context for understanding what we are detecting. Memory allows us to hear and see.

The brain then would be a whole community of neural networks. There would be networks that interpreted vectors coming from our senses, and networks that interpreted the output of other networks. How a perception was ultimately categorized would depend on the architecture of the system, that which a person was born with and that which was developed through experience. Some people's brains would tell them they had seen a UFO or an angel instead of a plane.

The work of people like Cooper and Anderson was leading to the idea that memory is a construct, not a videotape. Presented with the same event, different brains will pick different features to put into their memory structures, and they will build them in different ways. It depends, in part, on what is already in storage. Over the years, memories will get pushed together, the arrangements will shift and change. A single eyewitness report—whether in a scientific experiment or a criminal trial—should always be open to skepticism. It is only when the output of a number of different brains overlaps that we can take a kind of average and agree to call something reality. In the brain and in society, hordes of neural nets compete for the most plausible interpretation of the signals we receive from our world, matching them as best they can in a way that resonates with the past.

Something
Deeply Hidden

NEURAL NETS provided science with a layer of analysis that fell halfway between the abstractions of psychology and the numbing detail of neuronal chemistry. While Anderson spent the rest of the 1970s establishing ties with psychologists, Cooper headed in the other direction, deep into the bedrock of neurophysiology. If artificial neural nets, with their Hebb synapses and their random wiring, were to be more than engineering curiosities, people needed to find proof that something like them actually exists in the brain. Do the connections of the brain really change strength as Hebb predicted, allowing networks of brain cells to settle into new configurations, the patterns we call memories? In those days, the excitement over long-term potentiation in the rat's hippocampus had not yet caught fire. For those who were interested in how the brain was molded by experience, the hottest area was the visual cortex of the cat. So, once again, Cooper went back to the basics, determined to learn a new field from scratch.

Beginning in the 1960s David Hubel and Torsten Wiesel had shown that cats have brain cells that seem to work like feature detectors. Some cells fire most rapidly when the cat is seeing a vertical line, some respond more strongly to horizontal lines. One of the great questions in neuroscience was where these detectors come from. Are animals born with the circuitry intact, hardwired, coded in the genes? Or does the circuitry develop after birth as the animal interacts with its world? In that case, DNA would just provide the rough specifications for making randomly wired neural networks. Then these malleable devices would mold themselves to the environment, using something like Hebbian conditioning to develop the detectors needed to make sense of the world.

It was the old nature versus nurture question, cast in a way that was open to experiment. To tease out an answer scientists subjected kittens to all kinds of strange childhoods. Some were raised for a few weeks entirely in the dark, some in environments consisting entirely of vertical or horizontal stripes. Some began life with one eyelid sewn shut or with an opaque contact lens.

What Hubel, Wiesel, and a number of other researchers discovered was that cats—and presumably other animals—start life with a few feature detectors already in place. But these finely tuned cells are overwhelmed by the number of untuned cells, those that respond with equal enthusiasm to horizontal lines, vertical lines, or anything in between. In addition, the handful of early detectors that do exist are monocular: They respond to one eye or the other. At birth the kitten doesn't seem to integrate these two separate inputs into a visual whole.

But during the next few weeks, new circuitry grows. The number of feature detectors drastically increases, and the cells begin responding to images from both optic nerves. The cat learns that its two eyes are looking at the same world. Most important, scientists found that this development is not automatic, it is not programmed in the genes. If a kitten is raised in darkness, the blossoming of new feature detectors doesn't occur. But if these deprived animals are suddenly exposed to light, feature detectors rapidly begin to bloom. A few hours of visual experience is all that it takes for the circuits to form.

The lesson from all this was that the brain needs a world to respond to. The kittens raised surrounded by only horizontal stripes developed horizontal detectors, but they never grew the circuitry needed to see vertical lines. In the kittens with one eyelid sewn shut, the good eye quickly took over the entire visual cortex; connections to the bad eye seemed to wither away. But if the sutures were reversed, the newly opened eye developed connections that crowded out the old, now dormant ones. It was surprising how plastic all this wiring was.

. . .

Cooper collects scientific anecdotes the way other people collect stamps or old newspapers. As he pored over the data on the feline visual cortex, he was reminded of a story about young Einstein's reaction to seeing his first compass, the needle swinging insistently north: "Something deeply hidden had to be behind things." What was the hidden machinery that would explain how kittens developed the neural structures that allowed them to see? This mass of interesting data begged for a theorist to make sense of it, to construct an order that would charge the isolated facts with direction like iron filings aligned in a magnetic field.

Neural networks were at this point little more than metaphors. They suggested how something like a brain—with millions of fairly dumb processors working in parallel—could do things like generalization. For all their simplifications the models weren't inconsistent with what neuroscientists knew about the brain. But there had been little effort to unite experiment and theory, to show that a network could actually explain a large body of neurophysiological data. Now Cooper had what seemed like the perfect opportunity.

In the feline brain, as in the models Cooper and Anderson had been playing with, a largely undifferentiated web of neurons organized itself into a piece of complex machinery. But there was one important difference: In the case of the visual system, no one had trained it, giving it pairs of vectors one by one. Left on its own it adapted itself to its environment. This was just the kind of unsupervised learning that Cooper wanted to understand.

He started working on his model in 1973 when he and his wife were spending the summer in Sussex, England. Menasche Nass stopped by to discuss some ideas that would lead to a paper on the visual cortex published in 1975. The next year, Michel Imbert of the College of France, one of the most prominent researchers in visual development, arranged for Cooper to be appointed Professor of the Foundation de France. This allowed him to visit France several times a year until about 1983. Working in Imbert's lab, he began discussing a theory of visual development with Yves Fregnac, P. E. Buissert, and Elie Bienenstock, who later became one of his graduate students at Brown.

The result of this collaboration was a network that would learn to see like a cat. Until then, most neural networks had learned by adjusting their synapses according to some variation of Hebb's rule: If two neurons tend to fire simultaneously, then increase the strength of the connection between them. But Cooper figured that there would also have to be a way to turn the synapses down: Kittens with one eyelid sewn shut lost neural connections.

People had played with this notion of anti-Hebbian learning since at least the early 1970s. But Cooper brought the idea to a new level of refinement. He proposed that when a postsynaptic cell is stimulated by a presynaptic cell, two very different things may happen. If the receiving cell is firing above a certain threshold when it receives the signal, then the synapse should, à la Hebb, increase in strength. But if the neuron receives a signal while it is less active, the synapse should decrease in strength. The circuit can be turned up and down.

To test the idea, Cooper and his students simulated neural networks. The nice thing about simulations is that they focus the mind, forcing a theorist to sharpen vague ideas into something specific enough to run on a computer. In the course of these experiments the scientists were confronted with a difficult problem: where to set the threshold between learning and antilearning. Just how active or inactive did the receiving neuron have to be for the synapse to be turned up or down? After considering a number of configurations, Cooper decided that for his model to work in a way that would explain the data, he needed what he called a sliding threshold. The point at which a synapse would respond to an incoming signal by turning itself either up or down would constantly change. The value, Cooper decided, would depend on the average activity of the postsynaptic cell.

This is not quite as complicated as it sounds. Imagine that a neuron—a feature detector—has a hundred other neurons feeding it signals, each through a different synaptic connection. Any ten of these presynaptic neurons firing at the same time will form a pattern: a vector representing a horizontal line, a vertical line, or some angle in between. Now at first the neuron receiving these signals is untrained— all lines look the same to it, any input pattern will cause it to fire with more or less equal vigor.

But neurons are unpredictable devices. Suppose that on a single occasion some random influence causes the neuron to respond just a bit more strongly to one of the input patterns, perhaps the one meaning *horizontal line*. Now suppose the input pattern is quickly repeated— the cat is still seeing the horizontal line. This time the neuron will be firing rapidly enough when it receives the signal to exceed the learning threshold. According to Cooper's rule, the ten synapses that carried the signal are all turned up. The next time the neuron gets the same pattern it will come through stronger synapses. The cell will fire even more vigorously, and all the synapses will be turned up again. A kind of positive feedback keeps raising the volume of the cluster of synapses that mean *horizontal line*.

This is where the sliding threshold comes in. Since the cell is now responding more vigorously to horizontal lines, its average output increases, and, according to Cooper's theory, the learning threshold rises. Now the other input patterns—vertical lines, forty-five-degree lines—are not strong enough to fire the cell to its new threshold, so those synapses are turned down. Step by step, the synapses carrying the horizontal signal strengthen; the others fade away. Once the cell is tipped, it leans more and more in that direction. Its tuning curve becomes narrower and narrower. It becomes a feature detector for horizontal lines.

For his model, Cooper designed a network of neurons that behaved according to these rules. At birth, it was almost a tabula rasa, with a few cells given a weak tendency to respond to lines with different orientations. When the system was given patterned input— the kind of signals that a kitten would get as it looked around its world—it developed feature detectors. At the same time, purely random input—the electrochemical noise the neurons would generate in a kitten growing up in the dark—caused the cells to become untuned, returning to the naïve state in which they would respond to anything. Cooper could run the model as though an eyelid were sutured shut; he could raise it in a world of vertical lines. In all these cases, the model developed in much the same manner as a kitten's visual cortex did. Cooper had invented a network that seemed to mimic a part of the brain.

Most important, he had developed something still rare in biology: an abstract theoretical model that seemed to explain many of the experimenters' hard-won data. Like all good theories, it was a marvel of economy: By drawing on the idea of Hebb synapses, Cooper suggested that the plasticity that occurred in the infant brain during development obeyed the same kind of rules that the adult brain used during learning. Cooper speculated that the mechanisms that first evolved as a means for adapting the visual system to its environment were later borrowed by the higher brain as a mechanism for associative memory. Our rich conceptual world, with its tissue of concepts linked to concepts, could be a by-product of this basic machinery. Plasticity was so valuable for survival that it was conserved and expanded, eventually resulting in brains that not only could learn to recognize regularities like horizontal lines but could see more abstract patterns and even develop concepts like learning and plasticity, engaging in this search for hidden orders.

There was no evidence at this point that synapses really worked the way that Cooper envisioned. But a theorist doesn't always have the luxury of working with only established facts. That is where imagination comes in. When Einstein cooked up the theory of special relativity, there was no proof that clocks ran slower at higher speeds, or that rulers contracted. Sometimes a theorist has to use his instincts to guess at the structure of the invisible.

· · ·

Over the years, the parallels between memory and the development of the infant brain continued to become more tightly drawn. By the mid-1980s it was natural to wonder whether the same chemicals were used in both kinds of plasticity. Scientists like Gary Lynch and John Larson had helped link NMDA receptors, those unique molecular switches, to long-term potentiation, showing that they unleash the biochemical reactions that seem to lead to Hebbian learning. At about the same time, Wolf Singer and his colleagues at the Max Planck Institute for Brain Research in Frankfurt found that APV, the chemical that blocks NMDA receptors and prevents LTP, also interferes with the development of the kitten's neural circuitry.

In recent years, Mark Bear, a student of Cooper's who had gone to Frankfurt as one of Singer's postdocs, returned to Brown University as an assistant professor. He began working with Cooper on a molecular explanation for synaptic weakening and strengthening. He and a graduate student, Serena Dudek (one of Lynch's undergraduate protégés), have found evidence that neurons contain two very separate biochemical mechanisms, one for learning and one for antilearning. While NMDA receptors let in calcium, the trigger for the biochemical pathways that lead to synaptic strengthening, other types of glutamate receptors—the so-called non-NMDA receptors—set off reactions that may account for synaptic weakening. Bear and Dudek are also working on a biochemical interpretation for Cooper's sliding threshold; sometimes a little bit of calcium is enough to set off the reactions that strengthen a synapse; sometimes a lot of calcium is required.

"Cooper is looking at synaptic modification from the neural network perspective down, and I'm looking at it from the biology up," Bear said one day in his office, his eyes darting around the room as he spoke. "And the question is, where is the common ground?"

If Bear is correct, Cooper's theory will gain a stronger foothold on reality. After all, engineers can design neural nets to perform all kinds of impressive feats. As McCulloch, Pitts, and Turing suggested, there are any number of machines that can be rigged to perform a computation like sorting vertical and horizontal lines. Starting from scratch, neural networks of many different configurations could be made to sprout feature detectors. But only some will have roots in biology.

"You can come up with a lot of solutions without ever having to consider the strategy that was used by the brain," Bear said. "That's great for neural network research, but it is useless for neural science. There are some people who say, Let's look to the brain for the answers. And in that respect I think Cooper was truly a pioneer. Because he was looking to the brain in the early 1970s—what's a reasonable way to perform synaptic modification in a neural network based on the biology?"

Cooper and Bear's theory is one of several competing to explain the development of the visual system. "This could be total fantasy," Bear said as he leafed through graphs and charts of experimental results. "I don't think it is, but it could be. But what is exciting is how a theory that can analyze the problem mathematically and formally gives us something to pin our results to and say, Well, the theory says the brain should work like this, so maybe. . . .

"One of the problems in biology that physicists complain about a lot is that it's atheoretical, just going from one hypothesis to the next without ever trying to put it all together. This theory puts it all together, and so far there really aren't any data that it cannot explain. We don't make any claims that we've solved the problem but that a solution appears within grasp."

So, step by step, Cooper moved deeper into the details of brain chemistry, following the opposite trajectory of Gary Lynch, who started with the neurobiology and worked his way up to networks.

The followers of these two approaches didn't always know what the others were doing. But viewed from a distance, it almost seemed as though two intellectual kingdoms were sending legions from opposite directions toward a common ground, converging on theories that would help bridge the ravine between brain and mind.

Knowledge Engineering

WHILE THE NEURAL NET researchers labored on in obscurity, a world unto themselves, the people in artificial intelligence were attracting a great deal of attention with their separate approach to understanding mind as machinery. They were also making a lot of money. In the early 1980s covers of newsmagazines heralded the coming of so-called expert systems packed with rules for diagnosing diseases or plotting winning financial and military strategies. The Pentagon continued to pour money into the field. Most university researchers were running their laboratories on military money, and many were soon supplementing their incomes by consulting for large computer companies and starting enterprises of their own. Venture capitalists began staking small ambitious companies with names like Teknowledge and Intellicorp, hoping that one would become the IBM (or the Fujitsu) of artificial intelligence. The companies seemed to spend as much energy on public relations as on research. Their slick press releases and advertisements held out the promise that any kind of expertise—chess playing, medical diagnosis, air-traffic control—could be boiled down to a finite number of rules that could be programmed into a machine. At the annual artificial intelligence conferences, the industrial exhibition halls were filled with booths where attractive young men and women hawked the wares of the new age of intelligent computers. Meanwhile, the researchers who had come to the conferences would convene to hear panels of experts bemoan all the premature publicity and make gloomy predictions about a coming "A.I. winter," when investors, journalists, and Defense Department money-givers would become so disappointed that artificial intelligence had not quickly lived up to its promise that they would abandon the field in droves. But in the meantime, the researchers, including many of those on the panels, seemed to enjoy the attention and the money. It was nice to be quoted in *Newsweek*, even nicer to own a piece of a start-up company about to go public.

Behind all the hype was a serious idea, the deep-seated notion that intelligence could be studied as a thing in itself—a set of programs that could be made to run on any Turing machine. While the claims of the entrepreneurs were exaggerated, it was truly impressive how far one could carry this idea of the mind as symbols manipulated according to a set of well-defined rules.

In most of the A.I. systems, the knowledge was all a priori. The memories were implanted from the start by the programmer—learning by brain surgery, some critics called it. But the most impressive of the programs could actually learn, to some degree, on their own—they had rules for making new rules. A program called Eurisko, invented at Stanford University by Douglas Lenat, taught itself to play a sophisticated space war game called Traveller and won the national championship two years in a row. For training, it played against itself, pitting one fleet of imaginary ships against another. By carrying on these simulated battles night after night, it gleaned principles that would make for a winning Traveller battalion. Eurisko was able to do this because it had been provided by its inventor with concepts like *meson gun*, *armor*, and *radiation damage* and rules like this: "If a concept proves occasionally useful but usually worthless, then try creating a new, more specialized version of it." In one case Eurisko applied this rule to itself, coming up with this variation: "When specializing a concept, don't narrow it too much—make sure the new version can still do all the good things the old one did."

But all of the programs were very limited in scope. Eurisko could be made to learn about Traveller and a few other well-delineated subjects like number theory and integrated-circuit design. Most of the programs didn't learn anything at all. They had to be given all their rules in advance. To make a salable system—an artificial expert in analyzing mortgage applications, for example—programmers (or, in industry jargon, "knowledge engineers") would interview human experts and try to capture their knowledge, both the explicit and the implicit, in the form of rules. The challenge was getting humans to articulate rules that had become so deeply embedded in their behavior that they were all but unconscious—hunches, intuition, whatever people like to call it. For any but the simplest disciplines this was an overwhelmingly difficult task. The A.I. companies hinted at the day when they would offer artificial knowledge engineers—expert systems whose expertise was in making expert systems—but this was a distant dream. Few of the companies even bothered trying to make actual expert systems to sell. Instead they sold do-it-yourself programs called shells. Companies could purchase these generic programs and make their own expert systems by plugging in the appropriate rules. Whether they would actually be able to come up with an adequate set of rules was another story, but for a few years shells were all the rage—every innovative company had to have one to play with.

But the market quickly became saturated, the software wasn't getting much better, and one by one the A.I. companies began to fall into financial trouble. Expert systems found a useful niche in some corporations. But the heralded revolution in intelligent machinery remained stuck in its infancy. The technology just wasn't ready to live up to the hype. It didn't help matters that some of the most talented people in the field seemed to be spending as much time worrying about raising capital and the intricacies of the over-the-counter stock market as they were in trying to solve the huge problems involved in simulating the mind. By the late 1980s, the A.I. winter had set in.

For those interested less in engineering than in A.I. as applied philosophy—using computers as a tool to study the algorithms of the mind—important research continued. As simplified and schematic as the A.I. programs were, the basic idea seemed sound: The mind works as though it were following rules, manipulating tokens as if in some extremely complicated, multidimensional board game. Take the problem of pronouncing English words—not understanding them but simply converting printed text into speech. Viewed abstractly, the problem is a matter of taking one string of symbols, letters, and converting it into another, phonemes. In principle a Turing machine exists that can do this. For any input string it can be programmed to spit out the right output string—one set of symbols can be mapped onto the other. The A.I. approach would be to interview linguists and start writing down rules about how each letter and combination of letters is pronounced, depending on its surroundings. The letter *t* is pronounced as in *talk* unless followed by an *i*, as in *caption*. Then the rules would have to be indexed so that they were quickly accessible—you wouldn't want to have to search through the entire list for the appropriate rule each time you encountered another letter. But no simple rules would tell you that *tough* is pronounced differently from *though*, *bough*, and *cough*; that *daughter* doesn't sound like *laughter*. So the system would also have to store exceptions. Choose the wrong rules, and *ghiti* could be pronounced like *fish*.

Wouldn't it be wonderful if instead of tediously assembling, plank by plank, this complex knowledge structure, you could make a learning machine and train it by giving it examples, letting it learn the rules on its own? Not a learning machine like Eurisko, which had to be provided at birth with an intricate set of symbols and learning rules, but a true child machine, an electronic tabula rasa? In the late 1980s, as A.I. was starting to lose some of its sheen, some frustrated researchers were starting to think again about neural nets. Why not start from the bottom up and take a system of artificial neurons, expose it to a world, and let it learn the symbols and the rules? As they started looking into the literature, some of these researchers were surprised to find that the field hadn't died with Frank Rosenblatt, that people like Cooper and Anderson had been quietly working away.

Another Change
in Fashion

IN RETROSPECT, there are all kinds of perfectly rational reasons for the neural net revival that began midway through the 1980s. A decade and a half of research in neurobiology and neurophysiology had uncovered a wealth of good reasons to believe that memory really does work by changing the strength of synapses. It helped to know that the modelers might actually be modeling something real. At the same time, cheaper, more powerful computers made it easier for researchers to simulate neural nets instead of building them. It was no longer even thinkable to solder together transistors and volume controls and electric motors. Rosenblatt and his followers simulated some of their early machines on the primitive computers available in those days. But the new computers allowed researchers to experiment with more layers, more neurons, a higher density of synaptic connections.

But maybe the most important factor was a whole new generation of researchers who hadn't read Minsky and Papert's book. Early in the decade, the medium became ripe for a change in fashion, one of those collective decisions that emerge from the interaction of thousands of people sensing something new in the air. Many of the younger A.I. scientists—those who were truly interested in where mind comes from and not how to sell expert systems—were tired of working in a field whose goals seemed so remote. Douglas Lenat, whose Eurisko program remains one of artificial intelligence's great accomplishments, gave up in frustration when faced with the difficulties of trying to design better and better learning programs. He turned instead to a mammoth programming project—attempting to codify the vast amount of commonsense knowledge a reasonably intelligent being needs to make sense of the world, and putting it into a general-purpose computer data base. This was learning by brain surgery practiced with a vengeance. It would take nothing less, he believed, to make intelligent machines. But other researchers hoped against hope that neural networks offered a quicker, more direct approach.

Some of the new wave of enthusiasts like Geoffrey Hinton came from computer science, while others like David Rumelhart and James McClelland came from psychology. One of the brightest stars, Terry Sejnowski, came from neuroscience by way of physics. He started out studying at Princeton under a physicist named John Hopfield. Like Cooper, Hopfield went on to become one of the most important neural net theorists. In fact, by the end of the decade Cooper was no longer so unusual a figure. The field was beginning to generate the kind of interdisciplinary excitement that surrounded molecular biology after World War II when the physicist Erwin Schrödinger wrote *What Is Life?* and inspired younger colleagues like Francis Crick to give up physics for biology. Having helped unveil the double helical structure of DNA, Crick himself took up neuroscience, hoping to do for mind what he had done for life. With his biting wit, he has played the role of gadfly in a neural network discussion group that began meeting in La Jolla, California, in the early 1980s. La Jolla, in fact, soon became the hotbed of the revival. For a while, Rumelhart and McClelland were on the faculty of the University of California at La Jolla. Patricia Churchland, who has become the field's resident philosopher, is there. Adjacent to the campus is the Salk Institute, where Crick and Sejnowski work. Just up the interstate toward Los Angeles is the Irvine campus, where Lynch and Granger are using neural nets to study memory in rats.

To some critics, there is something a little too California about neural nets with their emergent, synergistic properties bubbling up from below. But with its roots in psychology, computer science, mathematics, and biology, what began emanating from La Jolla had all the appearances of serious science. One of the first things this new wave of researchers did was to try to transcend the limitations laid down by Minsky and Papert for single-layer perceptrons. By using more powerful computers they could simulate networks with several layers. Of course this put them face to face with another problem Minsky and Papert had anticipated: If some layers are hidden, then how do you know which synapses to reward and punish during a training session? Maybe Hebbian conditioning was enough: Any synapse between simultaneously firing neurons would be automatically turned up, no matter how many layers deep it was. But perhaps more sophisticated learning rules were called for. Rumelhart and Hinton helped push the field forward a notch by developing another approach: a learning rule called back propagation. If, during training, a network gives the wrong response to some stimulus—it says B when it should say A—error signals are sent back to the neurons in the hidden layer. Then their synapses are readjusted. Although the rule is neurologically unrealistic—it would require signals to flow back across the synapses from dendrite to axon—it is not inconceivable that the brain could implement something like back propagation by using feedback loops. In any case, the hope was that the method would provide a stopgap, allowing experiments that would suggest more likely procedures.

In fact some of the most important accomplishments in the field were made by scientists who chose to ignore many of the neurobiological details, seeking more general rules of network behavior. In this sense, the field's new incarnation sometimes had more in common with artificial intelligence than it did with neuroscience. But many of the model builders were certain that uncovering general principles of network behavior would guide the search for how real neural networks remember and learn. Once scientists knew how a simplified artificial network did something like generalization or telling circles from squares, they could look for similar architectures in the brain: a network with so many layers, a certain density of connections, a preponderance of feedback loops. The network modelers would provide the anatomists and the physiologists with suggested maps for exploring the terrain, theories that could be supported or falsified by experiment.

One of the strongest proponents of this approach was John Hopfield. By making a few simplifying assumptions, he showed that the behavior of a neural network and its engrams could be described by a set of equations that can be graphed as a surface in multidimensional space. A network with one thousand neurons would require one thousand equations and one thousand dimensions. But the flavor of the idea can be captured in the three-dimensional space we're accustomed to visualizing. In Hopfield's theory, a single-layer network with its synapses set at arbitrary values can be imagined as a landscape with peaks and valleys, mountain ranges, foothills, ridges, basins, ravines. Turn various synapses up and down, and the terrain changes: Some valleys become deeper, others disappear, peaks become steeper or gentler; from the middle of a plane, a ridge begins to rise.

Each of the valleys is a memory, or what mathematicians call a basin of attraction. At its very bottom is a point described by three numbers: longitude, latitude, and altitude—a vector that could represent some piece of information. Multidimensional space, produced by a network with more neurons, would accommodate bigger vectors. The point at the bottom of a basin would require ten, one hundred, one thousand numbers to describe it. The vector would hold that much more information. Now suppose you want to retrieve one of these memories, but you only have a fragment of the vector. Her name was Nancy, she came from Long Island, her eyes were brown, you met her at some event during the last two years. Maybe this is the first three hundred or so of the one thousand numbers on the vector. The rest is a random jumble. Her last name might have been

Wright or White, maybe even Winter. Or was it winter when you met her? You're not really sure.

In Hopfield's landscape, this imperfect memory would be a point somewhere within the basin of attraction of the vector that represents Nancy. It would be somewhere on a slope leading to the bottom where the correct information lies. Retrieving the memory would be like placing a marble at this position, allowing it to roll down the hill until it comes to rest in the proper indentation. Her last name was Whitney. You met her at a party in New York. She said she worked at a bank.

All over the landscape are other memories with their own basins of attraction. If you had remembered only her first name, one tiny segment of the vector, the marble would have sat farther from the basin, perhaps on a ridge between two basins representing Nancys you had recently met. The marble would have an equal chance of rolling into either one. If the network were saturated with memories, the surface that described it would have basins that overlapped; the memories would be fuzzier. Your third-grade and fourth-grade teachers might be squished together. The name of the janitor at your elementary school was Mr. Whiting. Or Mr. White. Or maybe it was just that he had white hair. A small, shallow memory might be surrounded by wide, deep memories—stronger basins of attraction that make the weaker one all but inaccessible. One night you might surprise yourself by chancing upon one of these valleys in a random exploration. For no apparent reason, you vaguely remember an afternoon visit to a house in the woods that belonged to the mother of a childhood friend. The rest of the details are gone.

The metaphor of hills and valleys felt right, and Hopfield's mathematics made it precise. Learning—twiddling the dials on the brain's volume controls—was a way of producing a memory landscape. On one level, engrams could still be thought of as patterns of lit-up neurons, but by also visualizing them as basins in multidimensional space, scientists could get a better feel for the whole complex process of storing and retrieving memories. Hopfield showed that if you remove some of the neurons, the landscape would retain its general shape. Some valleys might become a little shallower; it would be more likely that a marble would roll the wrong way. But the lay of the land would not drastically change. The network was capable of graceful degradation.

None of these ideas was particularly surprising to people like Cooper and Anderson, who had been playing with similar concepts for years. But Hopfield's work increased the size of the audience by drawing stronger parallels with solid-state physics. Hopfield showed that memories could be thought of in terms familiar to physicists. A network retrieving a memory was equivalent to finding the lowest possible energy state in multidimensional space, the point of elevation where the imaginary marble would have the lowest amount of potential energy. Physicists were equally impressed when Hopfield showed that his network equations also described the behavior of a spin glass, that strange substance that had intrigued both physicists and neuroscientists for so many years.

But Hopfield nets had their problems. In trying to recall a memory, a marble could get stuck in a dip somewhere on the hill above the proper basin, a false minimum. Then it would never find the memory. Imagine trying to recall the name of the artist Paul Klee and becoming fixated on Gustav Klimt. You know that is not the artist you're trying to remember, but your brain is stuck. Borrowing an idea from past researchers like Little and Shaw, Sejnowski and Hinton showed that they could avoid the problem by injecting noise into the system, shaking the marble out of the niche where it was stuck. This noise could be thought of as heat, the random vibration of molecules. In honor of this parallel, they named their network the Boltzmann machine, after Ludwig Boltzmann, who founded the science of statistical thermodynamics.

. . .

Cooper, Anderson, and the neural network veterans must have felt a little annoyed at the excitement that accompanied Hopfield's work. Many researchers complained that the ideas in his paper had been kicking around for years. But, as Anderson put it, "bringing them all together, with detailed, clear, and powerful mathematical analysis, is creative work of the first order." The fact that it was published in the *Proceedings of the National Academy of Sciences*, the journal of the most prestigious scientific organization in the country, attracted the interest of scientists in a number of fields who had never before heard of neural nets. From there, the culture began to spread.

In paper after paper, all the old ideas were coming back into vogue.

As the field expanded and began to encroach on other domains, one could gauge the motion of the fronts by reading the table of contents for the proceedings of various scientific conferences. Starting in the mid-1980s a rising number of papers presented at the A.I. meetings were about neural networks. The same thing was happening at cognitive science conferences, which soon became an important meeting ground for neuroscientists, psychologists, and network modelers. Even the neuroscience conferences included more and more neural net papers each year. The neural net people also started their own group—two of them, in fact, one on each coast. In 1987 the San Diego chapter of the Institute of Electrical and Electronics Engineers began sponsoring annual conferences. In 1988 it drew sixteen hundred participants. The International Neural Network Society recruited more than twenty-five hundred members and held annual conferences of its own. Finally, toward the end of the decade, they began holding joint meetings, attracting computer scientists, psychologists, neuroscientists, and philosophers.

. . .

One of the most active proselytizers at these affairs was Hopfield's student, Terry Sejnowski, whose lecture circuit included the A.I. conferences, the cognitive science conferences, and even an appearance on the "Today" show, where he demonstrated one of the field's most engaging wonders: a neural network that learned to read aloud. During the performance, he regaled his audience with a tape recording of his invention, called NETtalk. "The beauty of this particular problem," he liked to say, "is that it speaks for itself."

The machine consisted of about three hundred artificial neurons arranged in three layers: an input layer, which read the letters, a few at a time; an output layer, which generated phonemes; and a middle, "hidden layer," which mediated between the other two. The neurons were connected with about eighteen thousand synapses. At first the synaptic weights were set at random, and NETtalk was a structureless, homogenized tabula rasa. Provided with a list of words, it babbled incomprehensibly. But some of its guesses were better than others. When it chanced upon a correct pronunciation, its behavior was reinforced by adjusting the strengths of the synapses according to the back-propagation scheme.

First the machine learned to distinguish between consonants and vowels, but it substituted the same consonant for all consonants and the same vowel for all vowels. *Na-na-na-na, da-da-da-da, me-me-me-me-me*, it would say, gradually cycling through different combinations. Sejnowski called that the babbling stage. But hours later, the babbling would be gone and the speech no longer continuous. There would be boundaries between groups of sounds. The machine had learned about words, pseudowords actually. At this stage the chunks of sound were unintelligible. But after a half day of training, the pronunciations became clearer and clearer until it could pronounce some one thousand words. In a week it could learn twenty thousand. Through trial and error, NETtalk evolved.

The machine was not provided with any of the menagerie of rules for how different letters are pronounced under different circumstances. No one had to tell it that *e* is silent at the end of some words—*nice, wade, grape*—but not in the case of *he* and *she*. Unlike an A.I. programmer, Sejnowski didn't start with symbols as the primitive units, the letters and phonemes that would be manipulated according to rules. His units were the connection strengths of the synapses.

"What's really happening here is that you're going back to the foundations of your subject and saying, Let's question some of the assumptions that were made," he said one day in his office at Johns Hopkins University, where he worked before moving to the Salk Institute. "Do we really start with the symbol as the primitive element? Maybe we should have something more fundamental than the symbol. Maybe we should create symbols out of structures that are more elementary. Like elementary particle physics, trying to make nucleons out of quarks. It turns out that if you use quarks, you can explain a lot more of the phenomenology that was observed at the level of interacting protons and neutrons. And similarly, we may be able to understand a lot more about the general properties of memory, of learning, of visual processing if we look at this subsymbolic level."

Once NETtalk evolved, it acted as though it knew rules of pronunciation. They became coded in the net's hidden layer, though at first Sejnowski had no idea where the rules were located or what they looked like. Like a biologist studying an organism, he was confronted with a system so complex that it defied any simple, immediate understanding. It took some difficult mathematical analysis to uncover this hidden knowledge and see what the engrams looked like. One by one, he would give the network a phoneme and note which neurons lit up. Then he would treat this representation as a vector in multidimensional space. Using something called cluster analysis, he could detect order among the engrams.

"It turned out to be very sensible," he explained. "The vowels are represented differently from the consonants. Things that sound similar are clustered together." The sound p is situated near b, hard c near k. Each of the vowels—a, e, i, o, u, and sometimes y—have regions of their own.

Whenever he repeated the experiment, the same coding scheme emerged. Though different groups of neurons were used each time to represent different sounds, the same patterns appeared. On his own, Sejnowski says, he never would have come up with this particular system. He invented NETtalk, but NETtalk invented the coding scheme.

. . .

One of the most perplexing things about neural networks continues to be this problem of just what the engrams look like. The representation schemes are usually not specified in advance by the inventor as they are in A.I. Once they emerge, they are difficult to interpret. Some researchers purposefully design networks in which the engrams are localized instead of distributed—the word *mother* would be represented by a single neuron lighting up instead of by a pattern spread throughout the network. But with distributed representations it is hard to know just what the activation of an individual neuron means. Say you have a network trained to recognize one hundred objects. Through careful study you notice that a single neuron lights up each time ten of these objects are presented. There is something all these things have in common, though it is not necessarily a quality we have a word for.

"It's a question of how you carve up the information in the first place," Geoffrey Hinton explained. "Think of a node in the network. Sometimes it will be on, sometimes it will be off. Suppose you made up a new term in the language—xness—and we say all these things have xness and all these things don't. And we define xness by when this unit is on. Then obviously this is a local representation of xness. So what people really mean when they talk about distributed representation isn't that you could never give any sense to what the individual units are doing. It's that the items in our normal language might not correspond to the individual units. It's kind of a subtle distinction."

The Man
in the Pineapple Shirt

CONSIDERING THE LOPSIDED RATIO between synapses and neurons, it makes sense to think of the brain as consisting mostly of wiring, with the important action taking place in the connections between the cells. Hebb himself spoke of his theory as a brand of "connectionism." And the leaders of the neural network revival started calling themselves the new connectionists. By 1988 they were feeling confident enough to meet the man they regarded as the enemy face to face. To give the keynote address at the second annual IEEE neural net conference in San Diego, they invited Marvin Minsky, who was diplomatically referred to in the society's press releases as both a "neural network pioneer," in deference to his work on SNARC in the early 1950s, and one of the field's strongest critics.

And so one Sunday in late July, Minsky found himself at a Sheraton hotel on San Diego's Harbor Island, where the conference was being held. A short man with a bald head and glasses, he was dressed for the occasion in a black Hawaiian shirt decorated with large pineapples. Pacing around the hallways outside the meeting rooms, he nervously chain-smoked cigarettes.

"For many years I've had a religion of not preparing talks," he said a bit later, agreeing to retire to the hotel lounge for a nonalcoholic drink. "I have a lot of slides. So before I talk I'm going to put them in groups and then see what happens. But when I give these unprepared talks I usually end up with a new idea. Some people don't like these, but most people do. It gives them a chance to see thinking in action."

This was clearly not someone suffering a crisis of confidence. Minsky is famous for his sarcasm, focused on anyone he disagrees with or considers his intellectual inferior. This, it sometimes seems, includes just about everyone outside the M.I.T. artificial intelligence lab. But what was he going to tell this crowd of young researchers who considered the whole A.I. movement something of an anachronism, if not an outright failure?

Recently a new edition of *Perceptrons* had been published to capitalize on the neural net boom. In a new epilogue, Minsky and Papert belittled the notion that there was any kind of revolution afoot. The target of their derision was a recently published neural net bible, *Parallel Distributed Processing*, edited by Rumelhart and McClelland. For every triumphant claim made by these new acolytes, Minsky and Papert had a counterclaim. They denied that the new multilayered networks represented any kind of quantum leap over the old perceptron, even with the new learning rules for adjusting the hidden layers of synapses. They believed that the biggest problem with networks was one of scaling: There was no reason to think that a small network capable of learning a fairly easy task could be scaled up to solve the kinds of harder problems that brains do. Many of the new networks, they pointed out, take tens of thousands of trials to learn a simple skill, like recognizing a small number of objects. Would training the network to recognize ten times as many things involve ten times as many trials or the number of trials squared or cubed? Even worse, perhaps the networks scaled up exponentially, so that multiplying the size of the body of material to be learned by a factor, n, meant raising the processing time to the nth power. Then the problem would be intractable. Learning something difficult with a neural net might take longer than the universe would exist.

This problem of computational complexity can affect any kind of program, whether a piece of A.I. software or a simulation of a neural net. We simply don't know enough to extrapolate much beyond the toy problems used in laboratory experiments. In truth, a little more humbleness was called for on both sides.

The next night Minsky stood on the speaker's platform in a packed hotel ballroom, still wearing the Hawaiian shirt, and said with a self-satisfied grin, "I'm not the devil." He conceded that a couple of sentences in *Perceptrons* now seemed overly pessimistic. But he insisted that the book itself was not what killed off neural nets. By the time it came out, he said, the field had already died. He and Papert were just trying to explain why. "That's a rosy picture," he admitted, "because I feel guilty about it." Then he went on to put the new developments into his own perspective. He explained why back propagation and the other new learning rules for multilayered networks were plagued with problems. And he made a point of attacking Sejnowski's NETtalk; he had personally listened to the tape and couldn't understand what the thing was saying. But all in all, this wasn't the broadside that his audience had expected. He seemed very much in the mood for reconciliation, on his own terms of course.

And so he described his vision for a model of the brain that would embrace both the neural net and the artificial intelligence approaches. Perhaps it is not really necessary to scale up the networks, he proposed. Perhaps it is enough to have networks that just do toy problems. The brain would consist of a population of very simple nets not much more complex than the ones the neural net modelers were playing with. But if they were all put together and allowed to communicate in some kind of code, a more complex system might emerge, one capable of humanlike intelligence. He calls this idea the "society of mind" and he had recently described it in a book. The book didn't specifically mention neural nets. Minsky called the little processors agents, and they could just as easily be thought of as subroutines in a very complex computer program.

"If you look at the brain with a microscope, you can see as you move around probably at least a hundred different kinds of architectures," he said on another occasion during the conference. "You look at the cortex, which is a pretty uniform organ, and in one place you'll see that it has six layers, and the neurons in the third layer each run in two directions, and you look a centimeter away and you'll see that the layers are a little bit different. Instead of branching into two things, the neurons in a certain area will branch into a big circular pattern and go a certain distance and stop. And what I hope the neural net people will discover in the next decade is how these slightly different architectures are good at different problems. And then the society of mind issue is this: Suppose you know a hundred kinds of low-level learning machines, each of which has different abilities and speeds and so forth. How do you make a management structure in which some of them are good at learning how to manage how others learn? And that's what a brain is. It's got maybe three hundred kinds of neural nets, and some of them are specialized for controlling the inputs and outputs to others, some of them are specialized for controlling how the others learn. Some of them are specialized to act as short-term memories, so if you interrupt one neural net it can store what it's got so far somewhere else and attend to something else. Otherwise you could never solve a hard problem.

"I'm really pissed off that people say, How come you didn't talk about neural nets in the book? These people have no imagination, because I talk about brain cells all the time. The book is about how you'd use a lot of different kinds of neural nets to make something smart. So the entertaining thing about this conference is that people come and say, Are you going to give a talk against neural nets? They're trying to make an adversarial situation. But I'm trying to tell them how to apply A.I. concepts to organize the nets."

A cynic might accuse Minsky of engaging in a bit of revisionism. After all, he hadn't really said anything new. The idea that the brain consists of a single, homogeneous network instead of a whole community of different architectures had been abandoned long ago. But many listeners took a more charitable view. In the end, he was offering a way for the two sides to join forces. Clearly neither had the whole answer. The network theorists could study what kinds of neural nets were best for which tasks. But the A.I. people would need to figure out how the networks talked to one another. How could you arrange a variety of different kinds of nets into a society that thinks like a brain? And what is the code that all these agents would use to communicate? Finding this architecture and this language—a set of symbols and the rules used to manipulate them—would be very much in the spirit of traditional artificial intelligence.

· · ·

In fact the neural net movement was already beginning to resemble A.I., though not in the way Minsky had in mind. At the conference in San Diego the exhibition hall was filled with companies hawking networks that could learn to read handwritten letters, for example, or recognize objects on an assembly line. In one exhibit a neural network robot learned to balance a broom by its handle. Many of the companies were trying to sell neural network development systems the way the A.I. companies had tried to sell expert system shells: do-it-yourself kits instead of real programs. Some people had even taken to agonizing over all the hype and publicity and worrying that they were setting themselves up for a fall. If neural net computers didn't quickly succeed in the marketplace, the number of customers wanting software shells might plummet. The neural net industry would follow A.I.'s sad life cycle.

Of all the companies displaying their wares, the one that attracted the most attention was Nestor Inc. Standing aside computer terminals, enthusiastic men and women demonstrated programs that could learn to recognize signatures on checks or evaluate mortgage requests after being trained with examples of what were considered good and bad applications. As in the early days of commercial A.I., it had become an article of faith that success depended on finding customers in the financial industry.

While the other companies were at most a few years old, Leon Cooper had started Nestor in 1975 with a colleague at Brown, Charles Elbaum. Now it occupied a suite of offices in a converted warehouse near the Providence waterfront. For years the company remained a very low-key operation. But riding the new interest in neural nets, it had become one of the leaders in the industry. Some of the ideas Cooper and his colleagues had discovered to explain how brains work were being incorporated into products.

Other researchers were also venturing forth from the world of pure theory. Gary Lynch and Richard Granger were working with a company that wanted to inscribe their entire olfactory network onto a single silicon chip instead of simulating it with software. The idea was to make computer memories that could categorize information. Again, the parallels with A.I. in the early 1980s and neural networks in the beginning of the 1990s were a little unsettling: University researchers who weren't involved in start-up companies were consulting for corporations. The Defense Department was studying whether to kick in some large-scale financing.

And so it seemed that the neural net people were on the verge of realizing an old dream. The fruits of engineering would pay for the research that would help us finally understand our own brains. But as with biotechnology, commercialization quickly led to conflicts between the needs of science and the needs of industry. When professors start companies, ideas once freely traded can become proprietary information. Jim Anderson found this very disturbing.

"People don't talk as much as they used to," he said. "And you really see that with Cooper. He's always been a little bit secretive, but he doesn't really talk about what he's doing now. I didn't even know Nestor existed for many, many years. I'd heard rumors about it, but I was fascinated to see if Cooper would actually mention it within my hearing. And as far as I remember, he never really did—except when I met him at the Nestor booth at a neural net meeting two years ago. That's how Cooper operates. He's kind of quiet.

"Back in the seventies and early eighties before Nestor became more commercial, he was much more open about discussing ideas. His company has had the effect of decoupling him from Brown in some respects. He's still around but not influential in the way he was. We don't talk as much. I kind of regret that—he's a very bright guy, he's one of the more impressive people I've met. But you lose that in this commercialization. People just clam up. They get involved with serious finances and nondisclosure pacts and all that. The whole thing about academics is that you're supposed to talk about anything you're doing, and when other stuff starts to get in the way—"

He broke off his thought in mid-sentence. "I don't have a company. I'm just not the type, I think." Still, he does consult sometimes for companies like Texas Instruments—recently, he said, on a Pentagon contract to use neural nets to recognize radar signals. But he said one of the guiding rules for the project was that all the techniques that were discovered be published and put into the public domain.

The Price of an Idea

ONE MORNING IN 1989 during the mania over the notorious Pons-Fleischmann experiments, in which a scientist at the University of Utah dominated the news with claims that he and a British colleague had solved the world's energy problems by creating fusion in a jar, Cooper arrived at Nestor, his head bubbling with ideas. He didn't quite dare to believe that cold fusion was real. But confronted with the mere possibility, his mind automatically began trying to construct a theory that would explain such a phenomenon, real or imagined.

"This happens to me every morning of my life," he said. "I just sit there at home with my *New York Times* and my big pot of tea, and after I have enough caffeine in me I can just feel my brain going from a barely conscious level to this high pitch, as though I've taken a drug. I'm suddenly enormously awake and very manic, as you can see. Ideas tumble out—almost all to be discarded by noon, unfortunately. But if I can focus on something where I really know the facts, where I really understand the problem, that's when something might happen.

"Here at Nestor I've worked on many practical little problems," he said. "I love solving problems like that. One can't constantly think about the large questions. You need something to do day by day. Certain things are just too massive to get inside our heads. The large problems must be broken into pieces and left there with the hope that something will happen. Occasionally it does. If you focus on something that's small enough, you can occasionally crack it. When you're excited about it, then it's not hard to concentrate. It's hard to think about anything else."

In offices all around him, programmers sat at consoles trying to embody some of the more down-to-earth ideas in the form of network simulations. Others demonstrated products to potential customers or discussed financial strategies. It was a very different atmosphere from a university research lab.

"We use a large number of architectures, but we can only talk about some of them," Cooper said. "It's very difficult to protect intellectual property. One choice is to keep everything secret, which to a certain extent we tried to do. And we were criticized because nobody knew what we were doing. As our patents began to come through, we decided to go public with some specific networks." But some of the knowledge must remain secret, he said.

"It's unfortunate that it is so difficult to protect intellectual property. This is not only damaging to the individuals and institutions that create this property, but it is damaging to the United States vis-à-vis our trading partners, since a reasonable part of our national product is intellectual property—research, ideas, software, et cetera. Copyrights give protection only against plagiarism. Patents are difficult to obtain and protect. And to protect know-how one has to keep it secret.

"To me it's a fundamental flaw. Why is it that intellectual property is protected for a fixed number of years, by patent, for example, whereas if you own a piece of real estate you have it forever?

"If you can arrange it so that a writer of a song gets a royalty related to the number of times that song is played on the radio, you can arrange it so that a person who has an idea gets paid—it can be done. We live in a market economy. People accept the idea that if you buy something, you pay for it. You pay for coffee, you pay for cups, you pay the rent, et cetera. But you don't pay for ideas. So what happens is that the people and the institutions that produce ideas—scientists like myself and the universities, for example—are at an incredible economic disadvantage. We're always begging for money. Support for research stands somewhere between investment for the future and charity, the first items to be cut where budgets are tight.

"Now all this is perhaps a bit grandiose. But present realities are that the best way to capture the commercial benefits of one's own ideas is to exploit them commercially oneself."

Cooper was obviously tired of hearing about the dangers of academics going commercial. As a professional theorist, a spinner of ideas, he was determined not to let go of his creations. One was reminded of painters who, from time to time, band together to lobby Congress for laws giving them some kind of control over their works and royalties, even after they are sold and resold and resold again. Who knows where one of Cooper's ideas might turn up, embedded deep in a program long after he and his colleagues are gone? As he tried to show in his physics textbook, science is not disembodied knowledge. It is important to remember where ideas come from: individual human brains.

. . .

After almost half a century, counting from the publication of McCulloch and Pitts's paper in 1943, the neural net field is still in its infancy. It is far too soon to know whose ideas will survive.

"Things have changed dramatically," Cooper said. "Physicists now have taken neural networks to heart. They fill the journals with articles—I don't want to make any comments about *that*. Biologists have begun to accept the legitimacy of these ideas. When we first talked about synaptic modification as being the basis of memory storage and learning, it was regarded as possible but unlikely. Nowadays it is universally accepted that something like this is going on. The only issue is how much, where, and what is the specific molecular basis? So whether the idea turns out to be right or wrong, attitudes and language have changed completely. We no longer talk about whether it would be possible. We talk about which receptors, which channels, which protein kinases—that's a big change.

"Many neurobiologists still don't accept these ideas. But the new generation is beginning to think this way; the older generation is— well, they do retire."

For the time being much of neuroscience remains resistant to the idea of the professional theorist, someone to help decide which data to focus on, which to throw out. Neurobiology is still data-rich and theory-poor. It will help to forge on, gathering more experimental information. But a grand structure that would explain the mind biologically might still require strange, unforeseeable notions.

"To understand how we learn and remember or where thinking, consciousness, and self-awareness come from may require a leap as great as that taken by Galileo when he ignored air resistance," Cooper said, "or Newton when he extended gravity from the earth's surface to the moon, or Einstein when he redefined time. It may be that various deep properties will arise in a relatively straightforward manner as collective properties of simple systems—or it may be that some very profound change of point of view will be required."

These ideas won't necessarily be lying there as patterns to be culled from the data. Imagination will be required.

"What Einstein did was to redefine the notion of the clock," Cooper said. "That was the key underlying idea of the special theory of relativity. He redefined time. The reason people find that so difficult to accept is that his notion of time differs from our own psychological notion of time. So you tell a person about Einstein's clock, and they say, That's all very nice, but it's not really time. That makes it almost necessary to conclude that Einstein's clock is an *invention*. So then you might say, Why should we use this bizarre clock instead of the one we're so familiar with? The reason is very simple. If you want to organize our experience with nature the way the physicist sees it and not just in the little limited realm we see here, if you use ordinary clocks you twist yourself into contortions. But if you use Einstein's clocks you have stunning clarity."

In trying to explain how difficult a problem neuroscience still faces, Cooper likes to tell a story. Suppose you came here from another planet. You have no eyes, no ears, just infrared sensors to help you get around. You notice that an object is thrown on your doorstep every morning. But you are not equipped with the concept *newspaper*. You subject this strange artifact to physical and chemical analysis. You weigh it everyday and see that it goes from thin to fat in seven-day cycles. You analyze the ratio of black to white and find that it is fairly constant. You note that the chemical composition of the paper sometimes changes. But in understanding what a newspaper is, much of that turns out to be irrelevant. Will you, the alien, ever make the leap and somehow realize that on the surface of the paper are rows and rows of tiny markings, that they cluster into patterns that carry information? And, if you are someday driven to make this radical hypothesis, is there any hope that you will learn to read the thing? "The problem we're working on," Cooper said, "may not be any easier than that."

"To say that science is logical is like saying that a painting is paint," he said on another occasion. "A painter uses paint to create an image on a surface, and a scientist uses logic to create a structure. But the insights about which way to go and what to look at and what not to look at are obviously not logical procedures."

SECOND INTERLUDE

"God Is
a Tinkerer"

Of all the people involved in research on the brain and mind, the last group to get excited about neural nets has been the neuroscientists. When John Hopfield spoke about network modeling at the Society for Neuroscience meeting in Toronto in 1988, the auditorium was filled with listeners. But as he explained the fundamentals of the creed—that it was not necessary for the models to be realistic in every detail, that a simulation didn't need to include things like ion channels—a few listeners could be heard grumbling about the audacity of this interloper from physics trying to tell them how to run their show. How could these modelers hope to contribute to neuroscience when they insisted on glossing over the details? Why, some of their simulations even had synapses that transmitted information in both directions!

But the biologists, physiologists, and anatomists had their own conceptual problems. In neuroscience there is such a deluge of uninterpreted data that it is impossible to know which is really important. Nowhere is this clearer than at the Society for Neuroscience conferences, which attract more than ten thousand visitors a year. On a typical morning a scientist is faced with a choice of more than a dozen "slide talks" ranging from the neurochemistry of hunger and thirst to "Cyclic AMP and Phosphatidyl Inositol in Serotonin Facilitation in Crayfish." The specialists sitting in on one discussion will often understand only a few of the presentations, and they might very well be baffled by the session going on in the next room. While these groups meet on the second level of the convention center—all of these facilities seem to have been stamped out by the same architect—downstairs in an immense auditorium row upon row of posters describe more than five hundred other scientific developments. Visitors browsing through the labyrinth of bulletin boards can pause to consider evidence that "electric cues elicit sonic courtship in the mormyrid fish *Pollimyrus isidori*" or ponder whether gerbils with inflicted brain damage use "retinal image size to estimate distance in a jumping task." In recent years, so many of these microdevelopments have emerged that posters put up for each morning's session are taken down at noon to be replaced with hundreds more. The conferences last a week. Midway through these marathons, many participants have reached such a level of saturation that they are most likely to be found in the hotel bar.

Which of these data are important? If forced to start from the very bottom and work their way up, the theorists would have to consider an overwhelming amount of possible juxtapositions of facts, most of which would turn out to be beside the point or wrong. As they say in A.I., the search space would be too large.

"It's as if you're trying to understand mechanics and you don't know whether the colors of objects are really crucial properties or not," Geoffrey Hinton said. "It turns out that color is more or less irrelevant to mechanics. But it's not quite irrelevant. You know those little things that spin around in the light inside a glass jar? They spin around because one side's white and the other side's black. So it's not quite irrelevant but *almost*.

"If you were really unlucky, one of those things in a jar might be the first thing you saw. But you don't know in advance which properties are irrelevant and which aren't. The sciences that have really made a lot of progress have done so by making use of an immense amount of idealization, by ignoring lots and lots of properties that aren't involved." By tinkering with models researchers might find it easier to tell which details are worth concentrating on.

In autumn 1988, at a conference in Cambridge, Massachusetts, sponsored by *Nature* magazine and audaciously called "How the Brain Works," some younger researchers like Sejnowski, Hinton, and Richard Morris—the man who collaborated with Lynch on the swimming pool experiment with amnesiac rats—came together with such greats of neuroscience as Francis Crick, David Hubel, and Max Cowan to discuss the neural network craze. The philosopher Patricia Churchland was also there. During one session Crick attacked one of Hopfield's colleagues, David Tank of Bell Laboratories, who had presented a paper on using networks to study the problem of speech recognition.

This was not neuroscience, Crick sarcastically declared; it was engineering.

Scientists hate being called engineers, and many of Tank's colleagues leaped to his defense. Crick had raised a touchy point. In his memoirs, *What Mad Pursuit*, published a year later, he wrote, "I cannot help thinking that so many of the 'models' of the brain that are inflicted on us are mainly produced because their authors love playing with computers and writing computer programs and are simply carried away when a program produces a pretty result. They hardly seem to care whether the brain actually uses the devices incorporated in their 'model.' "

Crick was not objecting to the importance of models in biology. He and James Watson had used metal rods and plates cut to represent the lengths of atomic bonds to put together their double helical model of DNA. Computers simply provided a way to build more intricate and dynamic structures. But Crick wasn't sure that the neural nets were rooted firmly enough in biology. To Crick, it seemed that they were being developed more for their own sakes, as exercises in creating thinking machinery.

The essence of theory-building lies in ignoring distracting details. But at what point do your models become fantasies? With a computer simulation, the question is more subtle. A model might operate correctly, translating letters into phonemes, for example, but was that because it demonstrated something fundamental that could be applied to brains? Or was it just an interesting toy, a mesmerizing piece of engineering? The most exciting part of the conference was a panel discussion on just where to draw the line between science and engineering. A compressed version of the dialogue captures some of the flavor of the debate:

Richard Morris: I'm reminded of a slide that [the psychologist] Richard Gregory often shows. It's a picture of a tram that instead of having wheels has legs, and there are people in this thing and it sort of makes its way along the world by sticking out its legs and moving along. Clearly, in thinking about a method of transport it's failed to discover wheels, which would be much more efficient. Now why I'm thinking of that is because some of the neural network models seem to me to be trying to tackle essentially engineering problems, trying to discover efficient ways of solving problems, say of transport, but

200 · IN THE PALACES OF MEMORY

are then discovering things which really have never even been discovered in evolution, such as wheels. Can we draw a sharp division between when neural network modeling is doing an engineering task, which really doesn't necessarily have anything to do with the nervous system, and when it's modeling something that might really be there?

Max Cowan: Models are so far removed from biological reality and so difficult to assess and evaluate that we really are in danger at this stage of recapitulating the history of psychology, which treated the nervous system as irrelevant as long as you were understanding the phenomena.

Patricia Churchland: If you don't have modeling, then in a way you have this wonderful collection of data that threatens to become a clutter. A lot of what we do, and a lot of what our brain does, depends on what is going on in circuits, small circuits and larger circuits. So however neurobiologically unrealistic the models might be to start with, if they can coevolve with the experiments then it seems to me you have a prayer of addressing what it is that's going on in the systems. If you don't do that, I don't know how on earth you're ever going to get from the level of the single cell to the level of the system.

Francis Crick: Yes, but, Pat, why do you have to start with models which at the first possible test are so unrealistic?

Churchland: Well, I think partly because the system is so horrendously complicated that you have to have some idea of what *could* be a way to do something like learning with very simple units—

Crick: The point at issue is not whether we should have simple models but whether we should have simple models with obviously incorrect features in them. The first test you do, they're bound to fail because of the assumptions. If we look for neurons, we can't possibly find the ones that they postulate. There is such an infinity of models that unless we have our feet down to earth by testing some of the features, we won't be able to thread our way through this terrible jungle of models.

David Hubel: My emotional bias against modeling, if any, has always been that an experimentalist doesn't want to end up being told what experiments to do by a theoretician and then, having done them, be told what they mean. We can mold a science any way we want, and I would find it a real pity if neurobiology ended up like theoretical particle physics, with ninety people doing one experiment costing five hundred thousand dollars, all because the theoreticians told them to do that.

I think we should do the modeling. These people have to earn a living, and they will whether we like it or not. That's perfectly all right. But the experimentation has to go on. I would prefer if the world were built in such a way that the experiments were done and the models were done by roughly the same people. The models, to my mind, have suffered more than anything else from a lack of feel for biology by people who have never done experiments. For some reason in neurobiology, it's terribly important to get a feel for things. You can't express in papers very well how things work, apparently.

Geoffrey Hinton: As modelers we'd be quite happy to constrain ourselves more to the details of neurobiology, if we understood why they were there. But if you always slavishly constrain yourselves to all of the details that you can possibly model without understanding why they are there, you'll get less understanding of what's going on.

Terry Sejnowski: Models can give you ideas, but they can't give you answers. Even bad models can help. Let me give you a few examples. The earliest model of electricity was the hydraulic model. It was wrong—Francis would agree with that—but it was useful. Or let's take the Bohr model of the atom. Bohr had the simpleminded idea of electrons as little planets going around the hydrogen atom, and he had a rule about quantization and, lo and behold, he predicted the hydrogen lines. He predicted them for the wrong reason. Electrons are not planets. The reason why the Bohr model of the atom was useful was that it was a stepping stone to the next level. It gave some confidence that this idea of quantization was a good one to pursue. And that's really what the value of a model is, to tell you you're on the right track. What you really need are constant interactions.

We can now begin to think in concrete terms. Something concrete focuses the mind in a way that any number of equations or abstract word theories will never capture. And that gives us something that we can put in our hands, we can hold, we can look at—and we can get ideas from looking at things. That's part of the way we think as human beings.

If one is really going to come down to it, let's ask about nature. I mean, how did we get here? It's taken us hundreds of millions of years to get here, and we've come up, in a sense, through a lot of different partial solutions that changed over the millennia. I mean, let's face it: God is not a scientist, God is an engineer. He builds things. In fact he builds creatures of immense diversity. What I don't find sitting around are scientific theories produced by God.

We have to start thinking in terms of an engineering approach to the problem as well as the scientific part. The scientific part is going to help us guide our ideas and lead us along the correct path. But I think that some of the essential ground rules are going to come from building.

Here is a quote from someone, and I want you to guess who actually produced this quote: "God is a hacker." I won't say who it was, but he knows who it was.

Crick: The person who produced the slogan "God is a hacker" is myself. I was very pleased with this slogan. I think I produced it at dinner at the Churchlands'. But it does have one fault, which was pointed out by a friend of mine. A hacker does have some purpose even if he does go on adding bits here and bits there. Evolution doesn't have a purpose. So I've stopped using the slogan. I'm trying now to think of a better one. François Jacob said evolution is a tinkerer.

. . .

So on one point, anyway, Crick and the network modelers would agree: The brain itself is the result of nature's engineering, a device created through the trials and errors of evolution. Somewhere along the way, brains developed the ability to construct the elegant over-simplifications we call theories. Part genetic, part cultural, this skill was obviously important to our survival; theory building, like fire building, was fruitful and multiplied.

But how close can we really come with our theories to under-standing the universes inside and outside our heads? Have we any cause to believe that the brain nature has equipped us with is up to the task? The naïve view of science—and, for that matter, of the human experience—is that we are information processors with legs, wander-ing through the world gathering data, building pictures in our heads. But as the pictures are put into place, they act as lenses and filters, distorting what we see, determining what we can think. The same questions about computer simulations of the brain apply to all of our theories and mental constructs. How tight is the fit between theory and reality? Where do you draw the line between the subjective and the objective? With problems like these, we move from the fringes of science into philosophy.

PART THREE

The End
of Philosophy

The totality of our so-called knowledge or beliefs, from the most casual matters of geography and history to the profoundest laws of atomic physics or even of pure mathematics and logic, is a man-made fabric which impinges on experience only along the edges.
—W. V. O. QUINE

Spooky
Stuff

IN 1969, while she was teaching philosophy at the University of Manitoba, Patricia Churchland sensed that there was a serious gap in her education. She had spent four years earning master's and doctoral degrees in philosophy, specializing in the nature of mental representations, but she had never seen a human brain. In fact, after a frustrating stint at Oxford University, she had learned that most of her colleagues had no interest at all in neurons and synapses. For that matter, they didn't have much interest in science. Instead they searched for what they called "a priori truths," principles of the mind that were supposed to be self-evident and accessible only through contemplation. They assumed that empirical knowledge, that which came through the senses, was too misleading to be the source of truth.

To Churchland, a practical woman who grew up in the wilds of western Canada, this seemed like the purest kind of nonsense. Ever since a high school biology teacher had tried to assure her that people are alive because they are animated by an inexplicable life-force, she had been suspicious of what she called "spooky stuff," phenomena that supposedly fell outside the sphere of science. At Oxford, the endless philosophical debates on the nature of mind and consciousness had left her cold; it seemed clear to her that without the input of the sciences no resolution to these problems was in sight. If she ever hoped to understand the mind, she had come to believe, she would have to learn everything she could about the brain. And so at the age of twenty-six, having just moved to Winnipeg to begin her career as a philosopher, she decided to enroll in classes part time at the local medical school. She still remembers the smell of the formaldehyde, and it brings with it a rush of memories.

"We would have lectures three times a week, and one morning of lab," she recalled as she sat in her office at the University of California in La Jolla. "Usually, it was dissection. Then about the third week, after we had done some microscope work, looking at neurons, we got a brain. Everybody got a human brain. They came in these rather large Tupperware pots. And what you had to do was take it apart and find all the major pieces. It was really quite thrilling. You look at this thing, and you realize this is what makes a human go, and in fact this is what made some particular, real human, who was loved and went to school—this is what made him work. And, of course, it's essential to do a dissection because otherwise you can't figure out where the thalamus is or where the amygdala is—because it is a 3-D problem—and looking at a 2-D projection in a book, you never really get it.

"It was just a time of my life when I was high all the time because I was discovering things that made sense to me, and I could see how neuroscience could tell us about the mind. I would teach philosophy, and then I'd get in my old beat-up car and roar downtown in the thirty-below weather to where the medical school was." After teaching something very abstract like logic, she would go see what a real logic machine looked like.

Churchland didn't discover any new secrets about brain function. But she came away from the experience convinced that the answers to the great philosophical questions—what is the mind, how does it represent knowledge—lay somewhere in those Tupperware pots among the tangles of neurons. To put it in philosophical jargon, she was sure more than ever that she was a materialist, who believes that everything is made of matter and energy, as well as a reductionist, who believes that mental states and brain states are one and the same. Since those days in medical school she has committed herself to forming bridges between neuroscience and philosophy. She believes that the neuroscientists have a few things to learn about the nature of philosophical problems and the place of science in the big scheme of Western thought. But she is resigned to the possibility that neuroscience will someday subsume philosophy. Naturally, this position makes her rather unpopular with some of her colleagues. The belief in a priori truths, accessible not through experience but through pure reason, is still widespread in philosophy.

"There are some people who just think that there are deep truths that can be fathomed only by pure reason, and they're going to operate on the assumption that these ideas are in Plato's heaven and go for it," Churchland said. "It's a good idea to have a lot of different hypotheses pursued independently, but on the other hand I do find some of the projects in philosophy a bit odd because they seem to assume this rather Platonic realm of knowledge.

"Part of the difficulty is that if the profession begins to recognize that the empirical data need to be brought in when you decide some of these questions, then the argument goes that it's going to ruin the profession—philosophy is not going to exist anymore. It's going to be absorbed into psychology and into neuroscience, except for some things that really can't be approached yet but may be someday, like ethics."

But to preserve the privileged status of philosophy would require one to believe in a knowledge accessible through some route other than the five senses. And that would mean believing in spooky stuff.

"I want to know and understand, and what I care about is the truth," Churchland said. "And if your discipline's boundaries get shifted a little bit in the course of finding things out, then it doesn't seem to me that that should signify. I mean it used to be the case that the nature of space and time and motion and fire and life and the nature of the heavens—these were all philosophical questions. So I think that philosophy is bound to look different in fifty years because between them, psychology and neuroscience are going to teach us so much about the things we thought were uniquely and forever philosophical—about knowledge, about consciousness, about free will."

. . .

As the twentieth century enters its final decade, a number of thinkers have become seized with the millenarian spirit. Writers have declared the end of history, the end of nature, and the end of art. It seems anticlimactic to announce the end of anything else. But as neuroscience comes closer to explaining more about the brain and mind, some scientists, and even a few philosophers, are wondering if philosophy itself could be approaching its denouement.

Biology, geology, physics, astronomy—most of what we now call science—once fell under the heading of natural philosophy. But slowly, over the centuries, the study of the physical world has been taken over by people who call themselves scientists. Of course science too is a philosophy, one of any number of world views that compete to explain the mysteries of existence. But among all the competitors, science has been so persuasive and successful in allowing us to predict and manipulate nature that its tenets are all but taken for granted; for most people they are not even subject to debate. Believers in science tacitly assume there is a world out there that can be known through the senses, that it is made of nothing but matter and energy and operates according to laws that are the same throughout the universe and throughout time.

In theory, scientists are supposed to be content with explaining the how of existence, leaving the ultimate question of why to the philosophers and theologians. But as they push toward a grand unification principle that would unite the four fundamental forces into one great superforce, scientists are trespassing on the area of thought philosophers call metaphysics—that which goes beyond mere physics, mere description. Einstein showed that matter and energy are interchangeable, and to some extent, so are space and time. In another act of unification, he showed that one of the four forces, gravity, arises from the curvature of space-time; aesthetics almost demands that the other three forces—electromagnetism and the strong and weak nuclear forces—also be conjured up from geometry. Then space, time, matter, and energy would all be united into some mathematically abstract unity, the great cosmic One. That would still leave some rather important questions, like why is there something instead of nothing, and what is the meaning of our own existence? This is where the line between physics and metaphysics is supposed to be drawn. But having come this far toward a universal explanation, few scientists are content to suddenly stop short. To explain why there is something instead of nothing, some physicists have embraced the anthropic principle, the notion that the universe is as it is because otherwise we would not be here to observe it. Others speculate about how quantum fluctuations in the vacuum (the nothingness) could have spawned the Big Bang.

When they step this far toward the edge, scientists are on as shaky ground as philosophers. Since their cosmic extrapolations are built from the same bricks as electromagnetic wave theory, quantum theory—the philosophies that give us things like television—the speculation assumes a certain amount of authority. But it is easy to get carried away. Like anyone seized with an ideology, some scientists will not be content until they have explained everything from beginning to end with the same principles. As the physicist Leon Lederman once said, the goal of grand unification is to come up with a "theory of everything" so succinct that the equations can be printed on a T-shirt.

To achieve this kind of closure, science would also have to train its sights inward and develop theories of what it means to theorize, and of the relation between our models and the world. Until recently, it was easy to argue that while science was beating out the philosophers in offering answers to most of the questions of ontology, the study of what exists, another huge philosophical preserve was left untrammeled: epistemology, the study of knowledge. How do we know what we know about atoms, galaxies, or even the people next door? It is through epistemology that philosophers discuss the nature of memory in its broadest terms: what is knowledge, and how is it represented in our heads?

For centuries philosophers have talked about the nature of mental structures. Kant called them schemata. Psychologists and artificial intelligence researchers began to encroach on this terrain with their scripts and frames, ways of arranging knowledge into useful packets. But it has only been in the last few years that these abstractions have begun to assume a more corporeal existence as scientists develop plausible theories about how memory structures might be acquired and kept inside the brain. A scientific theory of memory would go a long way toward explaining how we represent the world internally, carving it into the categories we call knowledge. Plato found the notion of concepts so puzzling that he proposed the existence of a parallel world where pure forms and ideas lie. How could there be an idea like *tree* when every tree on earth is different? The theories of synaptic modification and neural networks make the existence of concepts much less mysterious—treeness would be a kind of average taken by the brain as it is exposed to instance after instance of real trees.

And so epistemology too seems destined to be taken over by science. As physics moves toward an explanation of existence, neuroscience, psychology, and computer science are coming ever closer to explaining the very minds responsible for the unification. To whatever extent there is mental stuff—ideas, information, memories, scientific theories—it seems to consist of matter and energy interacting in an arena of space and time.

Philosophy's Hinterlands

PATRICIA SMITH CHURCHLAND is used to being out of the mainstream. A tall woman with blue eyes and blond hair, she was born in 1943 in a small farming town in the wilds of British Columbia. It was a poor community, its only resources the scenery and, as she recalls it, the conversation.

"My father was a most unusual person," she said. "He had grown up first on the prairies and later in a little tiny, tiny sort of two-horse town. And his school only went up to grade six, so he went to grade six, and that was it for school. Then he went to a slightly larger town and worked for a number of years as a printer's devil. And he just read everything he could get his hands on. So eventually he ended up being a small-town newspaper publisher and a farmer. He was just very smart and very knowledgeable, and he read all the right stuff. I guess he must have read garbage too, but he read *The Origin of Species*, for example, and he really understood what it was all about. And he read geology, auto mechanics, physics, and history, and he was just very curious about lots of things, so it meant that at home there was always this atmosphere of discussion. At the same time, he was a highly practical man. If he needed something done, he did it himself.

"Of course lots of our friends were people who were very religious, and so there would be after-dinner arguments about evolution and God and about heaven and politics. My father was very, very sharp and impossible to hornswoggle, so arguing with him was very useful. He was gentle, but he would always make sure that you were clear about where an argument was hanging together and where it wasn't.

"A lot of the farmers were somewhat similar to my dad. There were a lot of Englishmen who were very eccentric but well read and spoke their minds. They talked about other things too, like the irrigation system and which kind of spray to use. But intellectually, it was surprisingly rich given how deep in the outback we were. I think I really did grow up in what was a philosophical atmosphere, but I certainly didn't know it at the time. I just thought that was the way people carried on."

After high school she enrolled at the University of British Columbia in Vancouver. "It was just not thinkable, given the money, that students from my school went anywhere else. In fact it was mostly not thinkable that they went to college at all. And it wasn't, I'm sorry to say, a very good university, though I didn't know that at the time. I thought it was interesting and fun and exciting. And I did learn some philosophy, but when I got to the University of Pittsburgh for graduate school I realized there were people from City College in New York and other places who just knew so much more than I did. I mean, I didn't even know some of these things existed. It was sort of embarrassing."

By the time she got her bearings, she found that she was leaning in the direction of epistemology. "I was always interested in the problem of knowledge and representation and how we know and understand," she said. At the time she had no reason to believe that these weren't exclusively philosophical questions. After earning her master's degree in 1966, she began looking around for a good place to get a doctorate.

"I probably should have stayed at Pittsburgh. They really had a fabulous philosophy department. But for various reasons, most of which were personal, I decided that I really wanted to go abroad, and I thought that it would really be interesting to see Oxford. But once I got there, the things that I had learned at Pittsburgh began to undermine my confidence that the way things were done at Oxford was the way I wanted to do them."

At Pittsburgh she had felt somewhat out of her element, competing with more sophisticated students from better schools. But studying at Oxford was a completely alienating experience. She found herself surrounded by a faculty that had become steeped in the writings of J. L. Austin, who pursued something called ordinary language philosophy.

"The idea seemed to be that you could really come to understand things such as perception or reasoning or morals or what have you by analyzing the concepts that ordinary people use when they talk about those things," Churchland explained.

"If you wanted to know something about whether or not actions are caused by certain things like motives or desires or whatever, then instead of doing some actual experimental work, psychological or otherwise, you would just kind of analyze what we mean by *motive* and what we mean by *cause*. John Austin had done some work on *real* that was like this. Descartes had raised the skeptical problem of how we can be sure we're awake when actually we may be sleeping, and what we think is real is just a tremendously vivid dream. Austin took the view that if we analyzed the word *real* as it is ordinarily used, we could see that, of course, we can't be dreaming, that reality is what we all call reality. And that, I think, was a turning point for me, because Austin just seemed to miss the point. If you could solve the skeptical problem just by analyzing the word *real* as it is ordinarily used by telegraph operators and waitresses, that would be easy, but I really didn't think that was the way it should work. If you pressed analytic philosophers, they would say these are a priori truths that we're discovering. And I would say, How can they be a priori truths? Where would they come from?"

A few of those who weren't caught up in these linguistic games had turned to religion for ultimate answers.

"There was a group of philosophers who had some sort of attachment to the Catholic church. Elizabeth Anscombe, for example, was a convert to Catholicism and believed quite literally in most of the doctrines of the church. Now between these two groups, it meant that science as a way of finding out something about the nature of the mind was not taken very seriously, either because people thought that you had to do this linguistic analysis stuff first or because they thought that there was really a species of truth that science could never hope to capture but that could be captured by deeply reflective and possibly spiritual people. I suspect there are still a lot of people who think that, quite apart from Oxford. My feeling is that most of the profession probably still operates on some sort of assumption like that. At least half of my colleagues do."

Churchland stuck with her interest in the nature of mental representations, taking something of a psychological approach. For her dissertation she decided to write about how the mind goes about turning desires into actions. She was specifically interested in how desires were represented in the mind and how science could go about finding the structure of the representations. Since she wasn't interested in approaching these questions by meditating on the meaning of *desire* and *action*, and certainly not through an exegesis of Catholic theology, she had trouble finding a supervisor for her project. But she had one kindred spirit to discuss these matters with—her fiancé, a philosopher from the University of Toronto named Paul Churchland, whose specialty was the philosophy of science.

"About the time that I was worrying about these things, Paul came over to England, and we worked a lot of them out together," she said. But her dissertation was not very well received. She left discouraged but with a degree.

· · ·

After failing to get a job with Paul in Toronto, the two of them began applying to other Canadian universities. As it turned out, the job market was tight for husband-wife philosopher teams. Finally, a friend at the University of Manitoba in Winnipeg told them that the school was hiring. They visited the campus, gave talks, and were offered jobs in the philosophy department. Before moving there, they went to her family's farm and got married.

Academically, an appointment at the University of Manitoba was not exactly an impressive move. The philosophy department was all but unknown. Nice as it was, Winnipeg seemed an isolated outpost, surrounded by the Manitoba prairie. The nearest Canadian city of consequence, Toronto, was thirteen hundred miles away. "We thought it was going to be really terrible because it was so remote," Churchland said. But the very remoteness fostered a kind of looseness that offered certain advantages. Her department chair thought it was just fine if she wanted to dissect brains in a medical school lab instead of churning out papers for the philosophical journals.

"If I had been at a top-ranked university, it would have been very difficult," she said. "They would have said, Well, look, you've got to write a whole lot more papers before you're going to get promoted. But at the University of Manitoba—I mean, I don't want to run it down or anything because there are some wonderful people there, but the standards were much more relaxed. People really didn't care if I went to medical school. If I basically met my teaching requirements, then they didn't care what I did. And if I didn't publish anything, they really couldn't have cared less, especially since some of them weren't doing anything. They didn't want to look bad. You can't really say that's the way you want a university run, but the lack of pressure was wonderful. I mean, this whole thing could have bombed. I could have gone off to medical school and never really put anything together and said, Well, that's that, and gone back to teaching logic. But nobody was riding herd on me and saying, What's this going to come to and Is this really worthwhile and Where are the papers, We haven't seen any papers published for two or three years, What's going on?— nothing like that. So I had a kind of freedom that I think most people never have."

In addition to taking medical classes, she spent a lot of time at the school's neurology clinic seeing patients with brain damage. She also worked in a spinal cord laboratory and learned to do some simple experiments. All the while she discussed the philosophy of science with her husband, Paul. In 1982 and 1983, they spent a year at the Institute for Advanced Study in Princeton. While he wrote a book called *Matter and Consciousness*, a clear, unjargony introduction to the philosophy of mind, she worked on her own book, *Neurophilosophy*.

By the time the Churchlands moved to La Jolla in 1984 and took jobs at the University of California, she was putting the finishing touches on the book, which was published two years later. By straddling two disciplines with completely different perspectives on the brain and mind, she hoped to appeal both to philosophers who knew nothing about neuroscience and to neuroscientists who knew nothing about philosophy. It was ridiculous, she believed, that these two groups still didn't talk to each other. While the first part of the book consisted of a crash course in neuroscience, including the anatomy of the brain and the physiology of neurons, the second part put neuroscience into the bigger context of philosophy and explained things like the mind-body problem.

"The sustaining conviction of this book is that top-down strategies (as characteristic of philosophy, cognitive psychology, and artificial intelligence research) and bottom-up strategies (as characteristic of the neurosciences) for solving the mysteries of mind-brain function should not be pursued in icy isolation from one another," she wrote. "What is envisaged instead is a rich interanimation between the two, which can be expected to provoke a fruitful co-evolution of theories, models, and methods, where each informs, corrects, and inspires the other.

"For neuroscientists, a sense of how to get a grip on the big questions and of the appropriate overarching framework with which to pursue hands-on research is essential—essential, that is, if neuroscientists are not to lose themselves, sinking blissfully into the sweet, teeming minutiae, or inching with manful dedication down a dead-end warren. For philosophers, an understanding of what progress has been made in neuroscience is essential to sustain and constrain theories about such things as how representations relate to the world. . . . It is essential, that is, if philosophers are not to remain boxed within the narrow canyons of the commonsense conception of the world or to content themselves with heroically plumping up the pillows of decrepit dogma."

She knew that reconciling these two forces would be a difficult project. For the differences between neuroscience and philosophy were not just a matter of terminology and style. The split was deeply rooted in Western thought.

Celestial Navigation

FOR CENTURIES the history of ideas has been driven by a tension between rationalism and empiricism. The debate turns on the value of reason versus experience in learning truths about the world. In its purest form, rationalism holds that the only way to construct a picture of the world is to start with a handful of self-evident truths—the so-called a prioris—and use logic to deduce the laws of the universe. The problem, of course, is where do these truths come from, and how do you know them when you see them? The empiricists believe instead that everything that can be known has to come in through the senses. The problem here is what relationship do these sensory data have to the real world? Are there really red apples out there, or is the redness in our minds, a mere product of the nervous system? If we are really born as blank slates, as the empiricists say, then how do we organize the onslaught of data? Don't we need to be born already knowing about things like cause and effect and space and time?

Between these two extremes, Kant struck a middle ground: Our brains come equipped with some kind of innate knowledge, filters that help us sort the data into categories. Reality, then, would be a construct of the mind but one that was also rooted in some kind of outside world. With the rise of neuroscience, we see that there need not be anything mystical about innate knowledge; it could all be there in the way our brains are wired, a product of evolution.

As modern science developed in the eighteenth and nineteenth centuries, it maintained an empirical slant. A scientist might have conceded a few points to the Kantians: that red is as much a product of the nervous system as of the apple. But even if the brain did filter the data, hardly anyone in science saw that as a reason to question the validity of studying the world by gathering facts and looking for patterns—what philosophers call the process of induction. But for those who worried about such things, this presented some messy philosophical problems. The opposite approach to truth seeking, the rationalists' process of deduction, had the virtue of being neat and orderly: You started with things you knew (somehow) were true and used logic to derive other truths. But induction was extremely suspect. The classic objection went like this: You might see a hundred white swans and conclude that swans are white, but there is no way to know that the next one you see won't be black, overturning the theory. Everything science could say was tentative. There was no such thing as absolute empirical truth.

Most scientists were perfectly content with tentative knowledge. They assumed that science was a process. By constantly refining their models they would come closer and closer to the mark. Many philosophers also agreed that this was the best way to proceed. But they were troubled by the unruliness of the venture. Surely, there was some way to make induction more philosophically respectable. And so, early in the century, logical empiricism was born, a failed attempt to make science as airtight and unquestionable as possible.

Bertrand Russell, Rudolf Carnap, and the other logical empiricists believed they could find a solid base for science in sensory data so rudimentary that its truth could not be questioned—*red here now*. Then they could use the latest developments in symbolic logic to weave these facts into theories. Inductive knowledge would still be tentative—there was always the possibility of a black swan popping up—but logical empiricism seemed to provide a method to slowly put together the best obtainable model of the world. Most scientists have come to tacitly accept this description of their endeavor without thinking much about it.

But as scientists work away in blissful ignorance, philosophers keep finding serious problems with the methodology. The logical empiricists liked to think that they were being very hardheaded and commonsensical when they decided to root truth in raw sensory data. *Red here now*—who could doubt such a fundamental observation? The problem is that any observation is affected by the knowledge that is already in our heads. The Kantians talked about innate knowledge—the wired-in filters—as though everyone came equipped with the same set: space and time, cause and effect, and so on. But as we learn new things we see the world through different eyeglasses. To put it in contemporary terms, the settings of the synapses in our neural nets are constantly changing; as Leon Cooper noted, memory and perception are inseparable.

Churchland found a striking example of this when she began studying neuroscience. A student does not look into a microscope and instantly begin seeing dendrites, axons, synaptic vesicles, and so forth. The ability to see these things comes with the acquisition of theoretical knowledge. The theory allows us to see the facts. Churchland described the experience in her book:

"A beginning neuroscientist's first observation through a microscope may produce puzzlement—it may be difficult to know what is artifact and what is part of the cell. ('*That's* endoplasmic reticulum?') Theory informs observation, and after a short while it becomes hard not to see, say, the end bulb. Notice that if I see something as an end bulb, then I imply that a whole range of different properties will obtain: that it is at the end of an axon, that if tested it would be found to contain synaptic vesicles, that if we looked at it under an electron microscope we would see synapses, and so on. To apply the descriptive term 'end bulb' to what is seen in the microscope is not a sheer, naked observation; it implies an indefinite number of generalizations applicable to the object. This cascade of indefinitely many implications is a general feature of the observational application of any descriptive term, whether it is 'coyote,' 'red,' or 'synaptic vesicle.' "

If everything we observe is embedded in a background theory, a network of knowledge both scientific and commonsensical, then the matter of interpreting an experiment is far from trivial. Say that an experiment fails to confirm a theory's prediction. The standard view would be that the theory is wrong and must be discarded. But as the philosopher W. V. O. Quine showed in the 1950s and 1960s, the flaw could be anywhere within the network of theoretical knowledge we bring to bear in formulating the theory or in designing, performing, and interpreting the experiment. For that matter, the flaw could be in the theories used to make the instruments—the patch clamp, the voltmeter, the oscilloscope, the dyes—used in the experiment.

Even worse, the experiment can never tell us where in the network of meaning the discrepancy between the world and our idea of the world actually lies. For all the lip service they pay to the logical empiricist creed, scientists understand intuitively how much slack there is in the system. No scientist is going to abandon a theory on the basis of one experiment, even if it repeatedly fails to work as predicted. Rather, he or she will readjust some auxiliary hypothesis, modify the architecture a bit so that the theory still stands. Proof that there was no glutamate binding occurring in Gary Lynch's test tubes did not cause him to abandon the calpain hypothesis. He made an adjustment instead.

What keeps science from being a free-for-all is the agreement to make the changes conservatively. In the early 1980s astronomers were startled to find a quasar that seemed to be ejecting objects at a speed faster than light. The choice was (1) to overturn Einstein's theory of relativity, (2) concede that the quasar wasn't nearly as far away as its red shift would indicate (meaning that our expanding universe was vastly smaller and younger than believed) or (3) find some less disruptive way to explain this celestial discrepancy. While a few maverick astronomers opted for number 2, calling for a revision in the age of the universe (this would have been in line with an agenda they had been pursuing all along), no one seriously suggested throwing out Einstein. Finally, astronomers settled on a trivial explanation that reduced the observation to an optical illusion. The perturbation was absorbed within the structure of physics and has been forgotten.

A more radical form of this maneuvering can be seen in the arguments between the creationists and the evolutionists. According to one popular analysis, counting biblical generations reveals that the earth is eight thousand years old. But radioactive dating supports the more widespread scientific view that the planet is some four billion years old. In an attempt to beat the scientists at their own game, some creationists have written papers using arguments from nuclear physics to show that the dating is wrong; after all, it is based on the assumption that uranium, for example has decayed into lead at the same rate since time immemorial. But who is to say? There is no way creationists are going to abandon the hypothesis of biblical infallibility, any more than the physicists would have abandoned relativity. Better, they think, to trace the discrepancy somehow to a problem in quantum theory. Likewise, some creationists read the Bible as requiring a geocentric universe and have written long mathematical treatises reinterpreting astronomical data to show that it is the sun, stars, and planets that circle the earth—Ptolemy's epicycles revisited.

Quine was not using philosophy to reject science. His point was that there is no privileged position—in philosophy, heaven, or anywhere else—from which one can stand outside science and judge how well it matches the so-called real world. We are inextricably part of the systems we try to describe. Churchland found this idea exhilarating. Quine was implying that the aim of science and philosophy both should be to understand how people use their brains to organize experience.

"[The] human subject is accorded a certain experimentally controlled input—certain patterns of irradiation in assorted frequencies, for instance—and in the fullness of time the subject delivers as output a description of the three-dimensional external world and its history," he wrote. It was up to scientists and philosophers to study "the relation between the meagre input and the torrential output." The question of memory—how experience is turned into the maps we carry in our brains—was central to science and philosophy. In fact Quine could see no real difference between the two endeavors.

As he once put it, "Science is like a boat, which we rebuild plank by plank while staying afloat in it. The philosopher and the scientist are in the same boat."

Rocky Waters

SINCE SETTLING IN at La Jolla, Churchland has become a familiar part of the community of neural network researchers and neuroscientists who are gathered around the University of California and the Salk Institute. The time has come for grand theories of the mind, she believes, and network models seem like the obvious way to go. She has collaborated on some papers and begun writing a book with Terry Sejnowski, championing what they call computational neuroscience, and she has taught a course on consciousness with Francis Crick. While the neuroscientists seem happy enough to have a philosopher on board, most of her colleagues in philosophy remain on another ship. Or, as some would like to believe, standing firmly ashore.

Churchland knows she has no hope of convincing the dualists, who believe that the mind is a separate essence that cannot be understood in terms of the brain. Nor does she expect to win over philosophers like John Searle, who deny that the brain is a kind of computer, nor physicists like Roger Penrose, who believe that the mind will not be understood until we have a grand theory linking quantum mechanics with general relativity. From the point of view of Churchland and the mainstream of neuroscience, these people are in a separate intellectual universe. But her own universe has fractures as well. One of her most formidable opponents is Jerry Fodor, who is squarely on the side of artificial intelligence and cognitive science. In philosophy, Fodor is probably the best-known proponent of the idea that the mind is a formal system. Like chess, checkers, and other games, thinking consists of shuffling meaningless tokens according to well-defined rules. From this symbol manipulation, the mind arises synergistically, just as a picture on a video display emerges from the shuffling of a computer's 1s and 0s.

This idea, called functionalism, gives artificial intelligence its philosophical justification—it's not the makeup of the brain that is important, but what it does. There is nothing about functionalism that necessarily conflicts with the neural net agenda. The neural net people are simply interested in focusing on how the symbols bubble up from below as a by-product of the neural interactions; the A.I. people believe in concentrating on the symbols themselves. There seems like plenty of room for compromise.

But Fodor joins the most radical of the artificial intelligence enthusiasts, insisting that to understand the mind it *only* makes sense to take the top-down approach—to study the software, not the hardware. A program can be run on any number of different kinds of machines, serial or parallel, so what is the point in studying the physiology of the brain? You would never learn how word processing worked by examining the circuitry in a personal computer; you have to get at the programs. In a way, the functionalists are no more interested in neurons than the ordinary-language philosophers were.

To Churchland this kind of attitude is as silly as that of the neurobiologists who bury themselves so deeply in the data that they reject any attempts at theorizing. Sure, a program can run on any number of different computers. But it is also true that there are any number of different programs that might be used to perform the same task. Which ones does the brain use?

"The functionalists say you don't really need to know anything about the implementation because, after all, you're interested in the software," she said. "But we don't *know* what the program is. And it could be that your biggest clues are going to come from the brain's architecture, or at least you want to have constraints from the behavioral level matched off against constraints from the architectural level." By looking to psychology and neuroscience, we might be able to converge on a theory from both directions, a description of the kind of mental software that is most likely to run on the machines we have in our heads.

"The brain is like other biological things," she said. "It's not likely to do things in the obvious way. We keep finding time and time again in biology that something we never would have thought of is what the system actually does."

God is a tinkerer. The brain is not likely to be a well-designed biocomputer whose flowcharts can be neatly laid out, but what Churchland calls "a Rube Goldberg machine": an accretion of evolutionary tricks for translating sensory data into mental structures—memories—that help ensure survival.

What a
Number Is

WARREN McCULLOCH ASKED, "What is a number that a man may know it, and a man that he may know a number?" The answer, he believed, lay in neurology, and in that interface between mind and the world that Kant first talked about. Our whole science, including the ideas we use to understand the brain, is built on mathematics. But if mathematics is partly a function of our nervous systems, and we use mathematics to understand our nervous systems, everything seems in danger of collapsing into a tangle of solipsistic loops.

With matters like these, philosophers press close to the edge of what it is possible to know and to think about. It is easy to despair, as Wittgenstein did, that science and all philosophies are ultimately closed systems of circular truths. Since it is impossible to step outside of these tautological webs, there will always be things we cannot know, stories that cannot be told. The universe will always overwhelm the theory builders. We are trapped inside our bodies and our nervous systems. As Quine said, We can't get off the boat. The best we can do, it seems, is to understand the nature of the filters and the mental structures, the mechanisms we use to convert experience into memories.

Assuming that there is an objective world out there that works the same everywhere—at least from our perspective—what would it be like for other creatures with other nervous systems? Can we even assume that they would look at things as being embedded in space and time? Or is that being anthropomorphic?

"I think there are regularities of some kind in the universe," Churchland said. "And it might be that there are different ways of capturing those regularities or describing them. But our brains, given the organization that they have, might be predisposed to a certain way of describing the regularities. So that if we met a physicist on Arcturus 4 who has a very differently organized brain, his physics might look very different but it might be as successful as ours."

But what about mathematics? Here is one area where even the hard-core scientist tends to go Platonic, arguing that mathematics is indeed discovered, not invented, that there are regularities that seem to have some kind of independent existence. The fact that there seem to be mathematical truths and perfect circles and triangles—things that you just don't find lying around in the physical world—has long been used as a counterargument to empiricism. Even the logical empiricists reserved a special place for mathematical truth, believing that it was separate from the truths induced by the senses.

But as neuroscience continues to close in on philosophy, it offers a way to explain mathematics without invoking spooky stuff. A circle *is* a mental object, something that we never really experience in life. But it doesn't have to be pure mind stuff. Rather, it could be a kind of compromise between the structure of the world and the structure of our brains. We see a lot of things that come pretty close to being round. The brain averages them according to its wired-in notions of order to come up with the concept *circle*. The fact that we use certain concepts—*circle*, *triangle*—and not others could be an accident of evolution. People with migraines often report hallucinations of geometric objects; there seems to be something neurological about them. There may be other ways to cut up the world that we can't begin to imagine. With different brains, these other shapes, or whatever they might be, would seem just as fundamental.

"There may be something we can tendentiously call mathematics that the people on Arcturus 4 do, and it doesn't look at all like what we do," Churchland said. "But it gets them around, and they manage to build spaceships, and they manage to use fire and smelt metal. So I'm not a Platonist with regard to mathematics at all. I don't want to be stampeded by an argument from ignorance—namely, But what *else* would make the propositions of mathematics true? I don't know what else, but maybe something in the brain. I think that about mathematics, and I also think that about logic.

"If you say that, of course, it usually sends people absolutely crawling up the walls. But I don't really see why. If the only alternative is that there are truths in Plato's heaven, then it seems to me that the basic story has to be told in terms of the brain. I just don't quite see how it could be otherwise."

FINAL REFLECTIONS

W̲HEN THEY TEACH techniques of meditation, Zen masters tell their students to make their minds as still as pools of water—so quiet that their thoughts float by, uninterpreted, like reflections of clouds on a lake. But minds are not mirrors; they are interpreting all the time. As soon as an impression impinges onto consciousness, it ignites memories and memories of memories. Soon it is hard to untangle one from the other.

As I write this, I am sitting in my apartment in Brooklyn looking out the back window at a steeple. It is raining, and tires are hissing on wet pavement. Cutting through this soft noise are distant sounds of construction: a power saw whining through a board. Several days ago, I returned from a visit to New Mexico, where I grew up. It always impresses me how quickly the vast expanse of a vacation becomes distilled into a handful of memories, a few stories to tell. By the time you return to work, the whole sprawling adventure has become compressed so that it exists entirely within a microscopic compartment in your head. And already the fading has begun.

I have the same feeling when I go to a big show at the Metropolitan Museum of Art. You spend an hour or two struck by one image after another, submerged in the work of Georgia O'Keeffe, perhaps, or David Hockney. Then you reach the end of the show and are inevitably forced to enter a decompression chamber—a harshly lit room filled with postcards and posters, pale carbon copies of the paintings you have just seen. The visceral experience of standing in front of *The Grey Hills* or *Mount Fuji and Flowers* begins to seep away. People around you rush to buy souvenirs, canned, prepackaged memories. Should you succumb to the temptation?

Memory itself works that way. The first time you tell somebody about your trip to New Mexico or your trip to the museum, it has already become trivialized, like the paintings. Raw experience has been converted into a few set pieces. And it is the set pieces that are remembered.

For a while, at least, the images of New Mexico are still fresh. I remember a day driving with a friend from my parents' cabin in the Jemez Mountains to Bandelier National Monument. The woman at the entrance gate looks too old to be working there, in a job that college students take for their summer vacations. She asks where we are from. "New York," I say, but I quickly add that I used to live in Albuquerque. I remember feeling glad that I am driving a pickup truck with New Mexico plates. "New York City?" she asks. For an instant, I take this as friendliness, then realize that she is only after information to be written in a ledger.

We drive down the road toward the canyon, deciding not to stop at an overlook. A car in front of us with Iowa plates is driving annoyingly slowly. As we descend, slowly, into Frijoles Canyon, I explain to my friend that the creek below us runs into the Rio Grande and this ignites memories of a backpacking trip I took through there years and years ago. And so, as I write this, I remember remembering—experience two steps removed.

We reach the bottom, where the visitors center is, and the Iowa car pulls into the first available space in the crowded lot. I feel a flash of annoyance. We might have trouble parking. But the feeling disappears as another car backs out of a space ahead, right in front of the entrance to the center. Perfect. I remember experiencing a quick, uncontrollable sense of superiority, then feeling stupid for feeling triumphant over something so trivial—just because this car was from Iowa and drove too slowly. I can still picture the parking lot, but hazily. I think, but am not sure, that the Iowa car was light blue.

And so the memories continue. We have iced tea and a Diet Coke at the snack bar. Sitting on a bench in a courtyard, I watch a man with a black beard taking pictures around our heads. I can't figure out what he finds so interesting. Then I notice a hummingbird feeder. He's taking close-ups of hummingbirds.

There is a gap now in the sequence, a hole a few minutes long— the time it took us to leave the patio and walk through the visitors center to the trail head. (It is not a complete blank. I vaguely remember deciding to go to the trail immediately, bypassing the museum. And

I remember worrying briefly that my friend thought I was rushing her. She likes to spend a lot of time in museums.)

But the images on the trail, the archeological ruins, are still with me. First we pass the Stone Lions, a concrete reproduction of a small monument on a distant mesa that the canyon's old inhabitants, the Anasazi Indians, used as a shrine. Next is a kiva, and I look down into it wondering where the sipapu is—the "spirit hole" that, as legend has it, connects life on earth with the world below. We pass the bearded photographer at a ruin called the Round House. I remember thinking that he is taking too many pictures and will get home overloaded with images. We pass a family resting on a bench, the little girl complaining about how hot it is. Then we begin the ascent toward the cliff dwellings, pausing to rest in the shadows. How cool it feels.

We walk up the stone steps quickly to put distance between ourselves and the complaining family. We visit the Talus House, climb up ladders into caves. A British group is in front of us. A girl in a yellow bikini top is hogging the ladder. The bearded photographer appears again.

And so my mind wanders from memory to memory, through the gallery of images. At the next ruin, called the Long House, an old couple is pointing out petroglyphs, pictures carved in the rocks. They have Midwestern accents (I realize only as I write this that they might have been the people in the Iowa car). I can still picture some of the images the woman showed us—a bird, a funny man. But they are faded and jumbled.

Another mile or so of details are written in my brain. At a junction in the trail that leads to more cliff dwellings, a family tries to decide whether to continue or return to the parking lot. A girl with dark hair and a hat stays with her mother. Two sisters continue on. I can't remember what they looked like. There is a gap again of several minutes, and then we are climbing up a series of ladders to the top of a cliff and another kiva. I can barely remember some kids up there arguing over whether their camera still has film. The father reprimands one of them for climbing where he isn't supposed to and getting stuck on a ledge. More vivid is the view up the canyon: pine trees and beige-and-pink rock.

The memory of the return trip, along a separate trail by the creek, is just as specific. I remember stopping, feigning interest in a sign on the nature trail, as a loud pair of women pass. We read more signs. Or, I remember reading them but not what they said. In my mind's eye I see a flash of what one of the signs looked like, but the image is too shaky to get a purchase on. A tiny garter snake slithers across the trail. Farther down, I almost step on a diamondback rattler.

Remembering this is like being in a trance. I'm amazed at how many specifics, including a lot of junk, are still encoded in my head. After the hike we drive to a grocery store in Los Alamos to stock up before heading back to the cabin. I remember the exact route we took through town, where we parked in the lot. I remember walking to a trash can to throw away a cup before entering the store. I remember almost every item we bought. We spot a pile of avocados and notice the ripe ones are marked down to twenty-five cents. I retain a strong, almost visceral feeling of holding the rough green avocado, squeezing it to make sure it is not too mushy. I remember discussing which wine to buy, and I can picture the row of bottles, though I can't zoom in and read the labels.

How different this all is from my memories of another trip I took to Bandelier three years ago with someone else. I remember that we took the hike through the canyon. But our position in the parking lot, the way the sky looked, our fellow hikers—most of that has faded. I know we must have seen the Round House and the Talus House because they are on the trail. It is only logical. But this is inference, not memory. I do remember that we saw the Long House, because the experience was unusual enough to stick. The trail was blocked with a sawhorse, and there was a sign warning about rabid bats. We went anyway, and, sure enough, there were dead bats lying everywhere. I remember posing for a picture on a ladder. Or do I just remember looking at the photograph weeks later?

And before that trip was another one, six years ago. In this case, all I remember is the simple fact that we went to Bandelier, nothing about what we saw.

Each time, as I walk along the trail, neurons link up with neurons forming all these memories, too many to keep forever. So the structures erode; the details fade. The actual images—the initial sensory impressions—disappear or become consolidated into symbols. Round House, Talus House, Long House—they exist as tokens in the mind. Remembering *what* becomes remembering *that*.

Maybe one of the reasons the recent trip is so vivid is that I've been on that trail so many times. Through repetition, it has worn a groove in my brain. I can picture the journey because I have memorized the sequence of the sights. Bandelier has become for me like Matteo Ricci's memory palace.

If experience is an indication, these latest memories will soon fade. The path through the grocery store, picking up the avocado— all of that will be compressed into the simple fact that we went to the store. Even now, as I retrace the memory structure, it is disintegrating. And the pieces that are left are being rearranged. When I remembered the avocado, my mind jumped ahead to a later scene up at the cabin. I had finished washing the dishes and draining the sink when I found a pot on the stove with a few leftover pieces of asparagus dyeing the water green. In the store and at the sink, that feeling of vegetable greenness was so similar and strong that a link was forged, a bridge from one memory to the other. Now I can't think of one without the other.

Even as memories are being laid down, the brain is consolidating and rearranging. The smallest details are lost as soon as they are registered: Even an hour later, I wouldn't remember every step on the trail. And as they form, the new structures are affected by what is already in storage. The reproduction of the Stone Lions was immediately connected to a memory of the real Stone Lions, which I saw on a backpacking trip a full fifteen years ago. Another bridge was formed.

When put on the witness stand, we can swear before God that we will tell the truth, the whole truth, and nothing but the truth. But the best we can really do is read out what is left of our memories, recollections that have been inevitably altered by time.

How many times have I closed up the cabin at the end of a vacation? I clean the fireplace, close the flue, board up the windows, sweep the floor, strip the beds, climb down into the pump house and turn off the well. This time I actually remember each of these acts. If the cabin had burned down and I had to testify that I had indeed cleaned out the fireplace, I could recall the actual act of shoveling (the blade was bent and awkward to use), of leaving two burnt, obviously cold chunks of wood, of dumping the dead ashes outside the back porch. But if someone asked me about a visit three years ago, I could testify that, yes, I probably cleaned out the fireplace, but only because it is logical that I would.

. . .

Postscript: As I edit these final pages, six months after I wrote them, I find that I was right: Most of the memories are gone. Rereading this is almost like reading fiction, someone else's short story.

ACKNOWLEDGMENTS

THE IDEA FOR THIS BOOK came to me in 1985 when I attended a conference of the Cognitive Science Society in Irvine, California, and heard a fairly obscure scientist named Gary Lynch make the astonishing claim that he had actually observed the physical changes memory leaves inside the brain. Since then I have learned much more about this and related subjects from interviews and conversations with Lynch and a number of people: Daniel Alkon, James Anderson, Philip Anderson, Michel Baudry, Mark Bear, William Calvin, Patricia Churchland, Leon Cooper, Carl Cotman, Serena Dudek, Chris Gall, Richard Granger, William Greenough, Shelly Halpain, Geoffrey Hinton, Douglas Hofstadter, Eric Kandel, June Kinoshita, John Larson, Julie Lauterborn, Jerome Lettvin, John McCarthy, Jay McClelland, Marvin Minsky, Richard Morris, James Olds, Kathie Olsen, Aryeh Routtenberg, Robert Schrieffer, Larry Squire, Terry Sejnowski, Charles Stevens, and many others. I would like to thank them all for inspiring some of the ideas that have found their way into these pages. Many of the scientists mentioned here were very generous about sending me reams of journal articles they had written, guiding me through the most difficult ones. Several of them read sections of the manuscript, helping me weed out errors.

The book first began to take shape in 1987 as an article for the *New York Times Magazine* called "Memory: Learning How It Works"; I'd like to thank my editor, Randy Rothenberg, for helping me with the piece. Several other editors at the *Times*, including William Borders, Richard Flaste, Erik Eckholm, Katy Roberts, and Marvin Siegel, helped with my education by agreeing to let me cover a number of fascinating scientific conferences. My favorites were those held by the Society for Neuroscience, which does more to help science writers than any organization I know of; I'd especially like to thank Donald Price and Eileen O'Donnell.

My friends Richard Freedman, Alan Lappin, Douglas Maret, Nancy Maret, and Katie Rosenthal played an indispensable role by reading the manuscript. Finally, in the world of publishing, I'd like to thank my editor, Jonathan Segal, and my agent, Esther Newberg, for once again helping me turn an idea into a book.

NOTES

UNLESS OTHERWISE INDICATED, all quotes are from interviews I conducted between August 1985 and April 1990 in Irvine, La Jolla, Pittsburgh, Baltimore, New Orleans, San Diego, Toronto, Providence, New York, and Cambridge, Massachusetts. Material also comes from lectures and other presentations at the annual Society for Neuroscience meetings in New Orleans (1987) and Toronto (1988), the International Conference on Neural Networks in San Diego (1988), and *Nature* magazine's colloquium, "How the Brain Works," in Cambridge, Massachusetts (1988).

Little is more ephemeral than a scientific paper, especially in a rapidly changing field like neuroscience. Rather than list all the publications I read, I have selected the ones most likely to endure (depending on whose theories turn out to be closest to the mark). If anyone has developed a standard system of citing neuroscience papers, I haven't been able to figure out what it is. Instead of trying to impose formal rigor, I've relied on efficiency and common sense. Whenever possible I've saved space by listing review articles or books that summarize the results of many earlier papers. Full references to books listed below can be found in the bibliography, which follows this section.

PREFACE: *Invisible Palaces*

PAGE

ix The quote from Bellow is on page 102 of *The Bellarosa Connection*.

xiii "To everything that we wish to remember": Spence, *The Memory Palace of Matteo Ricci*, 2.

PRELUDE: *The Tower in the Jungle*

5 The gathering of tribes occurred at the Seventh Annual Conference of the Cognitive Science Society at the University of California at Irvine, August 15 to 17, 1985.

PART ONE: *Mucking Around in the Wetware*

9 The quote from Jacob is on page 274 of *The Statue Within*.

14 Lashley's paper "In Search of the Engram" was published in the *Symposium of the Society for Experimental Biology* 4 (1950) 454–80, and is reprinted in Anderson and Rosenfeld's collection, *Neurocomputing*.

15ff The early history of neuroscience, including the Lashley-Penfield controversy, is described in a number of books including Rose's *The Conscious Brain* and Changeux's *Neuronal Man*.

16 Penfield's experiments are summarized in his book *The Mystery of the Mind*.

17 "The astonishing aspect": quoted in Rosenfield, *The Invention of Memory*, 163.

22–3 Hebb's theory is described in his book *The Organization of Behavior*, which is excerpted in Anderson and Rosenfeld's *Neurocomputing*.

23 Kandel's early work is summarized in his article "Small Systems of Neurons" in *Scientific American*, September 1979, 66–76.

24 "built like an old Philco radio": This quote first appeared in Stephen H. Hall's article "Aplysia & Hermissenda: Two Snails Are Leading the Race to Trace the Molecules of Memory," *Science 85*, May 1985, 33–34.

29 Lynch's early research on regulation of attention: "Separable Forebrain Systems Controlling Different Manifestations of Spontaneous Activity," *Journal of Comparative and Physiological Psychology* 70 (1970): 48–59.

30 Lynch's work on recovery of function was later written up in Byron A. Campbell, Percy Ballantine II, and Lynch, "Hippocampal Control of Behavioral Arousal: Duration of Lesion Effects and Possible Interactions with Recovery after Frontal Cortical Damage," *Experimental Neurology* 33 (1971): 159–70.

31 Sprouting was first documented by Raisman in "Neural Plasticity in the Septal Nuclei of the Adult Rat," *Brain Research* 14 (1969): 25–48.

32 Lynch's sprouting experiment: Lynch, Dee Ann Matthews, Sarah Mosko, Thomas Parks, and Carl Cotman, "Induced Acetylcholinesterase-Rich Layer in Rat Dentate Gyrus Following Entorhinal Lesions," *Brain Research* 42 (1972): 311–18.

36 Bliss and Lømo reported their discovery in "Long-lasting Potentiation of Synaptic Transmission in the Dentate Area of the Unanesthetized Rabbit Following Stimulation of the Perforant Path," *Journal of Physiology (London)* 232 (1973): 357–74.

   LTP is specific to the synapses: T. Dunwiddie, D. Madison, and Lynch, "Synaptic Transmission Is Required for Initiation of Long-Term Potentiation," *Brain Research* 150 (1978): 413–17.

   Lynch's experiment linking LTP with the creation of synapses: Kevin Lee, F. Schottler, Michael Oliver, and Lynch, "Brief Bursts of High Frequency Stimulation Produce Two Types of Structural Changes in Rat Hippocampus," *Journal of Neurophysiology* 44 (1980): 247–58. A preliminary report appeared in *Experimental Neurology* 65 (1979): 478–80.

43 The development of the calpain hypothesis is described in dozens of papers written by Lynch with Baudry, Siman, Staubli, and others. They are summarized and cited in Lynch's book, *Synapses, Circuits, and the Beginnings of Memory.*

54 The maze experiment is described in Staubli, Baudry, and Lynch, "Leupeptin, A Thiol Proteinase Inhibitor, Causes a Selective Impairment of Maze Performance in Rats," *Behavioral and Neural Biology* 40 (1984): 58–69.

55 A review of research on declarative and procedural memory is provided by Larry R. Squire in "Mechanisms of Memory," *Science*, June 27, 1986, 1612–19.

  Lynch and Baudry's "The Biochemistry of Memory: A New and Specific Hypothesis" appeared in *Science*, June 8, 1984, 1057–63.

56 The editorial "Skeleton Key to Memory?" appeared in *Nature*, January 17, 1985, 178–79.

58 Greenough and Chang's experiment is described in their paper "Transient and Enduring Morphological Correlates of Synaptic Activity and Efficacy Change in the Rat Hippocampal Slice," *Brain Research* 309 (1984): 35–46. A review of Greenough's work on the relationship between learning and synaptic development is provided in Greenough, James E. Black, and Christopher S. Wallace, "Experience and Brain Development," *Child Development* 58 (1987): 539–59.

59 Biographical details on Kandel and a description of his work are provided in Allport's book, *Explorers of the Black Box.*

68 Kandel's later work on classical conditioning is summarized in Thomas W. Abrams and Kandel, "Is Contiguity Detection in Classical Conditioning a System or a Cellular Property?," *Trends in Neurosciences*, April 1988, 128–35.

72 "a surprisingly radical reductionist possibility": in Abrams and Kandel, "Contiguity Detection," 135.

74 Alkon and his work are described in Allport, *Explorers of the Black Box.* More recent developments are summarized in his article "Memory Storage and Neural Systems," *Scientific American*, July 1989, 42–50.

76 Routtenberg's work linking LTP with protein kinase C: Raymond F. Akers, David M. Lovinger, Patricia A. Colley, David J. Linden, and Routtenberg, "Translocation of Protein Kinase C Activity May Mediate Hippocampal Long-Term Potentiation," *Science*, February 7, 1986, 587–89.

80 McNaughton's experiment showing that LTP is stronger when several pathways are stimulated simultaneously is described in McNaughton, R. M. Douglas, and G. V. Goddard, "Synaptic Enhancement in Fascia

Dentata: Cooperativity Among Coactive Efferents," *Brain Research* 157 (1978): 277–93.

Larson's experiments showing that LTP is a two-step process are described in Larson and Lynch, "Induction of Synaptic Potentiation in Hippocampus by Patterned Stimulation Involves Two Events," *Science*, May 23, 1986, 985–88.

81 Theta rhythms emanating from the brains of exploring rats were reported by C. H. Vanderwolf in *Electroencephalography and Clinical Neurophysiology* 26 (1969): 407. Theta rhythms in hippocampal cells were reported by J. B. Ranck, Jr., in *Experimental Neurology* 41 (1973): 462. The possibility that the hippocampus is where spatial maps are stored is explored in J. O'Keefe and Lynn Nadel, *The Hippocampus as a Cognitive Map* (London: Oxford University Press, 1978).

The link between LTP, the theta rhythm, and NMDA receptors is further developed in Larson and Lynch, "Role of N-Methyl-D-Aspartate Receptors in the Induction of Synaptic Potentiation by Burst Stimulation Patterned After the Hippocampal Theta-Rhythm," *Brain Research* 441 (1988): 111–18.

82 The original link between LTP and NMDA receptors was reported by Collingridge, S. J. Kehl, and H. McLennan in "Excitatory Amino Acids in Synaptic Transmission in the Schaffer-Commissural Pathway of the Rat Hippocampus," *Journal of Physiology (London)* 334 (1983): 33–46. Other work on LTP and NMDA receptors, including research by Gustafsson, Wigström, Brown, and Bliss is exhaustively described in two special issues of *Trends in Neurosciences*, one on learning and memory (April 1988) and one on NMDA receptors (July 1987).

87 The swimming pool experiment: R. G. M. Morris, E. Anderson, Lynch, and Baudry, "Selective Impairment of Learning and Blockade of Long-Term Potentiation by an N-Methyl-D-Aspartate Receptor Antagonist, AP5," *Nature*, February 27, 1986, 774–76.

88 Squire's work on the amnesiac patient R. B.: Stuart Zola-Morgan, Squire, and David G. Amaral, "Human Amnesia and the Medial Temporal Region: Enduring Memory Impairment Following a Bilateral Lesion Limited to Field CA1 of the Hippocampus," *Journal of Neuroscience* 6 (1986): 2950–67.

88–9 The experiment Lynch was gloating over at the Toronto conference is described in Dominique Muller, Michel Joly, and Lynch, "Contributions of Quisqualate and NMDA Receptors to the Induction and Expression of LTP," *Science*, December 23, 1988, 1694–97. Other recent details of Lynch's calpain hypothesis can be found in *Society for Neuroscience Abstracts*, vols. 13, 14, and 15, which describe papers presented at the annual conferences in 1988, 1989, and 1990.

FIRST INTERLUDE: *A Brain in a Box*

91 Squire and Zola-Morgan describe their conclusions about the anatomy of declarative memory in "The Medial Temporal Lobe Memory System," *Science*, September 20, 1991, 1380–1386.

95 The conflicting reports on synaptic memory mechanisms were published in *Science*, June 29, 1990, 1603–05, 1619–24.

97 For details of neural Darwinism, see Rosenfield, *The Invention of Memory*, 156–95 and Edelman, *Neural Darwinism*.

103 Lynch and Granger's neural network: Lynch, Granger, Larson, and Baudry, "Cortical Encoding of Memory: Hypotheses Derived from Analysis and Simulation of Physiological Learning Rules in Anatomical Structures," in Nadel et al., *Neural Connections, Mental Computation*, 180–224. More recent details are in José Ambros-Ingerson, Granger, and Lynch, "Simulation of Paleocortex Performs Hierarchical Clustering," *Science*, March 16, 1990, 1344–48.

PART TWO: *The Memory Machine*

109 Wallace Stevens's poem is included in *Poems by Wallace Stevens: Selected and with an Introduction by Samuel French Morse* (New York: Vintage, 1959), 54–55.

113 Cooper's ideas on the role of the observer in quantum theory: "How Possible Becomes Actual in the Quantum Theory," *Proceedings of the American Philosophical Society* 120 (1976): 37–45.

114 Cooper on the AIDS virus: "Theory of an Immune System Retrovirus," *Proceedings of the National Academy of Sciences* 83 (1986): 9159–63.

118 Some of Cooper's recollections on his role in the discovery of the BCS theory come from his lecture "Origins of the Theory of Superconductivity," delivered at the H. Kammerlingh Onnes Symposium on the Origins of Applied Superconductivity, October 1, 1986. The theory itself is described in Bardeen, Cooper, and Schrieffer, "Theory of Superconductivity," *Physical Review* 106 (1957): 162–64.

123 "Man comes into the world": Cooper, *An Introduction to the Meaning and Structure of Physics*, short ed., 117.

"The gathering of facts": ibid., 120.

124 "a special breed of people": Hubel, "The Brain," *Scientific American*, September 1979, 48.

125 Early work on the analogies between ferromagnetism and mass neuronal behavior is described in Jack D. Cowan and David H. Sharp, "Neural Nets and Artificial Intelligence," *Daedalus*, Winter 1988, 85–121.

127 For a recent description of spin glasses, see Daniel L. Stein, "Spin Glasses," *Scientific American*, July 1989, 52–59.

129 Turing's machine is described in his paper "On Computable Numbers with an Application to the *Entscheidungsproblem*," *Proceedings of the London Mathematics Society*, 2nd series, 42 (1936): 230–65.

131 The Turing test is described in his essay "Computing Machinery and Intelligence," originally published in *Mind*, October 1950, 433–60.

132 McCulloch's essays are collected in the book *Embodiments of Mind*.

134 McCulloch and Pitts's "A Logical Calculus of the Ideas Immanent in Nervous Activity" was published in the *Bulletin of Mathematical Biophysics* 5 (1943): 115–33. It is reprinted in Anderson and Rosenfeld, *Neurocomputing*.

   "Pitts and I showed": in Donald H. Perkel, "Logical Neurons: The Enigmatic Legacy of Warren McCulloch," *Trends in Neurosciences*, January 1988, 10.

   "Anything that can be exhaustively and unambiguously described": in McCorduck, *Machines Who Think*, 65.

135 McCulloch and Pitts's "How We Know Universals" appeared in the *Bulletin of Mathematical Biophysics* 9 (1947): 127–47. It is reprinted in Anderson and Rosenfeld, *Neurocomputing*.

136 "an idea": McCulloch, "Why the Mind Is in the Head," in *Embodiments of Mind*, 84.

138 Bernstein's profile of Minsky appeared in *The New Yorker*, December 14, 1981.

139 The perceptron is described in Rosenblatt, "The Perceptron: A Probabalistic Model for Information Storage and Organization in the Brain," *Psychological Review* 65 (1958): 386–408, reprinted in Anderson and Rosenfeld, *Neurocomputing*. For a clear analysis of how it works, see Chapter 15 of Singh, *Great Ideas in Information Theory, Language and Cybernetics*.

141 "For the first time": in Rumelhart et al., *Parallel Distributed Processing*, vol. 1, 156.

143 "to dispel what we feared to be": Minsky and Papert, *Perceptrons*, 19–20.

144 "without scientific value": ibid., 4.

242 • *Notes for Part Two*

146 For a look at artificial intelligence, see my book *Machinery of the Mind*.

147 The work of Minsky's early students is described in his anthology, *Semantic Information Processing*.

147 The philosophical debate over artificial intelligence has been written about extensively. See, for example, chapter 13 of *Machinery of the Mind*. For a rousing defense of the A.I. viewpoint see Hofstadter, *Gödel, Escher, Bach*. For the opposite view, see Searle, *Minds, Brains, and Science*, and Dreyfus and Dreyfus, *Mind Over Machine*.

151 The Longuet-Higgins paper Cooper described: "Holographic Model of Temporary Recall," *Nature*, January 6, 1968, 104.

152 "along with a lot of half-baked speculation": Anderson and Rosenfeld, *Neurocomputing*, 193.

153 "Although work on Aplysia": Anderson and Rosenfeld, *Neurocomputing*, 284.

154 "All that we are"; "In the beginner's mind": The epigraphs appear in two of Anderson's early papers, "A Simple Neural Network Generating an Interactive Memory," *Mathematical Biosciences* (1972): 197–220, and Anderson, Jack W. Silverstein, Stephen A. Ritz, and Randall S. Jones, "Distinctive Features, Categorical Perception, and Probability Learning: Some Applications of a Neural Model," *Psychological Review* (1977): 413–51; both are in Anderson and Rosenfeld, *Neurocomputing*.

154 Anderson's linear associator is described most fully in his 1972 paper, "A Simple Neural Network."

159 Cooper's first network is described in "A Possible Organization of Animal Memory and Learning," *Proceedings of the Nobel Symposium on Collective Properties of Physical Systems*, 1973, 252–64, reprinted in Anderson and Rosenfeld, *Neurocomputing*.

160 "If the world is properly ordered": Cooper, "A Possible Organization," in Anderson and Rosenfeld, *Neurocomputing*, 201.

161 Winston's work is described in my book *Machinery of the Mind*, 89–91.

"a fabric of events and connections": Cooper, "A Possible Organization," 201.

"The system is placed in an environment": ibid., 196.

162 Papers by Kohonen, Little and Shaw, Grossberg, and other neural network pioneers are included in Anderson and Rosenfeld, *Neurocomputing*.

164 "Many ways to store": Cooper, "Source and Limits of Human Intellect," *Daedalus*, Spring 1980, 1–17.

168 The paper Nass and Cooper wrote: "A Theory for the Development of Feature Detecting Calls in Visual Cortex," *Biological Cybernetics* 19 (1975): 1–18.

Cooper's later work on visual system development is described in Cooper and Michel Imbert, "Seat of Memory: Brain Theory Meets Experiment in Visual Cortex," *The Sciences*, February 1981, and in Elie L. Bienenstock, Cooper, and Paul W. Munro, "Theory for the Development of Neuron Selectivity: Orientation Specificity and Binocular Interaction in Visual Cortex," *Journal of Neuroscience* 2 (1982): 32–48; the latter is reprinted in Anderson and Rosenfeld, *Neurocomputing*.

171 Cooper's work with Bear is described in their paper "Molecular Mechanisms for Synaptic Modification in the Visual Cortex: Interaction between Theory and Experiment," in M. Gluck and David Rumelhart, eds., *Neuroscience and Connectionist Theory* (in press). Also see Bear, Cooper, and Ford F. Ebner, "A Physiological Basis for a Theory of Synapse Modification," *Science*, July 3, 1987, 42–48, and Dudek and Bear, "A Biochemical Correlate of the Critical Period for Synaptic Modification in the Visual Cortex," *Science*, November 3, 1989, 673–75.

177ff Recent advances in neural networks by Hinton, Rumelhart, McClelland, Sejnowski, Hopfield, and others are described in papers collected in Anderson and Rosenfeld, *Neurocomputing*. The key developments include: Hopfield, "Neural Networks and Physical Systems with Emergent Collective Computational Abilities," *Proceedings of the National Academy of Sciences* 79 (1982): 2554–58; David H. Ackley, Hinton, and Sejnowski, "A Learning Algorithm for Boltzmann Machines," *Cognitive Science* 9 (1985): 147–69; Sejnowski and Rosenberg, "NETtalk: A Parallel Network That Learns to Read Aloud," a 1986 technical report from Johns Hopkins University; Rumelhart, Hinton, and Ronald J. Williams, "Learning Representations by Back-Propagating Errors," *Nature*, October 9, 1986, 533–36. Other important work appears in Rumelhart et al., *Parallel Distributed Processing*.

181 "bringing them all together": Anderson and Rosenfeld, *Neurocomputing*, 457.

SECOND INTERLUDE: *"God Is a Tinkerer"*

199 "I cannot help thinking": Crick, *What Mad Pursuit*, 115.

PART THREE: *The End of Philosophy*

205 Quine is quoted in A. J. Ayer, *Philosophy in the Twentieth Century*, 247–48.

209 The millenarian thinkers are Francis Fukuyama, "The End of History?," *The Public Interest*, Summer 1989; Bill McKibben, *The End of Nature* (New York: Random House, 1989), and Arthur C. Danto, who declared the end of art in his collection of essays *The State of the Art* (New York: Prentice Hall, 1987).

217 "The sustaining conviction": Churchland, *Neurophilosophy*, 3.

219 "A beginning neuroscientist's first observation": ibid., 268–69.

221 "[The] human subject is accorded": in Gardner, *The Mind's New Science*, 70–71.

222 "Science is like a boat": in Churchland, *Neurophilosophy*, 265.

222 For a spirited debate between Searle and the Churchlands, see Searle, "Is the Brain's Mind a Computer Program?" and Churchland and Churchland, "Could a Machine Think?" *Scientific American*, January 1990, 26–37. Penrose's point of view is described in his book, *The Emperor's New Mind*. For a critique, see George Johnson, "New Mind, No Clothes," *The Sciences*, July/August 1990, 44–49.

222-3 Fodor's objections to neural nets: Fodor and Zenon W. Pylyshyn, "Connectionism and Cognitive Architecture: A Critical Analysis," in Pinker and Mehler, *Connections and Symbols*, 3–71.

BIBLIOGRAPHY

Adelman, George, ed. *Encyclopedia of Neuroscience.* 2 vols. Boston: Birkhauser, 1987.

Allman, William F. *Apprentices of Wonder: Inside the Neural Network Revolution.* New York: Bantam, 1989.

Allport, Susan. *Explorers of the Black Box: The Search for the Cellular Basis of Memory.* New York: Norton, 1986.

Anderson, James A., and Edward Rosenfeld. *Neurocomputing: Foundations of Research.* Cambridge: MIT Press, 1988.

Ayer, A. J. *Philosophy in the Twentieth Century.* New York: Random House, 1982.

Bellow, Saul. *The Bellarosa Connection.* New York: Penguin, 1989.

Campbell, Jeremy. *The Improbable Machine: What the Upheavals in Artificial Intelligence Research Reveal About How the Mind Really Works.* New York: Simon & Schuster, 1989.

Changeux, Jean-Pierre. *Neuronal Man: The Biology of Mind.* New York: Pantheon, 1985.

Churchland, Patricia Smith. *Neurophilosophy: Toward a Unified Science of the Mind/ Brain.* Cambridge: MIT Press, 1986.

Churchland, Paul. *Matter and Consciousness: A Contemporary Introduction to the Philosophy of Mind.* rev. ed. Cambridge: MIT Press, 1988.

Cooper, Leon N. *An Introduction to the Meaning and Structure of Physics.* short ed. New York: Harper & Row, 1970.

Crick, Francis. *What Mad Pursuit: A Personal View of Scientific Discovery.* New York: Basic Books, 1988.

Davis, Philip J., and Reuben Hersh. *The Mathematical Experience.* Boston: Houghton Mifflin, 1982.

Diamond, Marian C., Arnold B. Scheibel, and Lawrence M. Elson. *The Human Brain Coloring Book.* New York: Barnes & Noble, 1985.

Dreyfus, Hubert L. and Dreyfus, Stuart E. *Mind Over Machine: The Power of Human Intuition and Expertise in the Era of the Computer.* paperback ed. New York: Free Press, 1988.

Duffy, Bruce. *The World As I Found It: A Novel.* New York: Ticknor & Fields, 1987.

Edelman, Gerald M. *Neural Darwinism: The Theory of Neuronal Group Selection.* New York: Basic, 1987.

Fodor, Jerry A., and Zenon W. Pylyshyn. "Connectionism and Cognitive

Architecture: A Critical Analysis." In *Connections and Symbols*, edited by Steven Pinker and Jacques Mehler. Cambridge: MIT Press, 1988.

Gardner, Howard. *The Mind's New Science: A History of the Cognitive Revolution*. New York: Basic Books, 1985.

Gjertsen, Derek. *Science and Philosophy: Past and Present*. London: Penguin, 1989.

Graubard, Stephen R., ed. *Daedalus*, special issue on Artificial Intelligence. Cambridge: American Academy of Arts and Sciences, Winter 1988.

Hebb, Donald O. *The Organization of Behavior*. New York: Wiley, 1949.

Hinton, Geoffrey, and James A. Anderson, eds., *Parallel Models of Associative Memory*. Hillsdale, N.J.: Lawrence Erlbaum Associates, 1981.

Hofstadter, Douglas. *Gödel, Escher, Bach: An Eternal Golden Braid*. New York: Basic Books, 1979.

———. *Metamagical Themas: Questing for the Essence of Mind and Pattern*. New York: Basic Books, 1984.

Jacob, François. *The Statue Within: An Autobiography*. New York: Basic Books, 1988.

Janik, Allan, and Stephen Toulmin. *Wittgenstein's Vienna*. New York: Simon & Schuster, 1973.

Johnson, George. *Machinery of the Mind: Inside the New Science of Artificial Intelligence*. New York: Times Books, 1986.

Kline, Morris. *Mathematics: The Loss of Certainty*. New York: Oxford University Press, 1980.

Luria, A. R. *The Mind of a Mnemonist: A Little Book About a Vast Memory*. Cambridge: Harvard University Press, 1968.

Lynch, Gary. *Synapses, Circuits, and the Beginnings of Memory*. Cambridge: MIT Press, 1986.

McCulloch, Warren S. *Embodiments of Mind*. Cambridge: MIT Press, 1965.

Minsky, Marvin, ed. *Semantic Information Processing*. Cambridge: MIT Press, 1968.

Minsky, Marvin. *The Society of Mind*. New York: Simon & Schuster, 1986.

Minsky, Marvin, and Seymour A. Papert. *Perceptrons: An Introduction to Computational Geometry*. expanded ed. Cambridge: MIT Press, 1988.

Monod, Jacques. *Chance and Necessity: An Essay on the Natural Philosophy of Modern Biology*. New York: Alfred A. Knopf, 1971.

Nadel, Lynn, Lynn A. Cooper, Peter Culicover, and R. Michael Harnish, eds. *Neural Connections, Mental Computation*. Cambridge: MIT Press, 1989.

Pagels, Heinz R. *The Dreams of Reason: The Computer and the Rise of the Sciences of Complexity*. New York: Simon & Schuster, 1988.

Penfield, Wilder. *The Mystery of the Mind*. Princeton: Princeton University Press, 1975.

Penrose, Roger. *The Emperor's New Mind: Concerning Computers, Minds, and the Laws of Physics*. Oxford: Oxford University Press, 1989.

Regis, Ed. *Who Got Einstein's Office?: Eccentricity and Genius at the Institute for Advanced Study*. Reading, Mass.: Addison-Wesley, 1987.

Rose, Steven. *The Conscious Brain*. rev. ed. New York: Paragon House, 1989.

Rosenfield, Israel. *The Invention of Memory: A New View of the Brain*. New York: Basic Books, 1988.

Rumelhart, David E., James L. McClelland, and the PDP Research Group. *Parallel Distributed Processing: Explorations in the Microstructure of Cognition*. 2 vols. Cambridge: MIT Press, 1986.

Searle, John. *Minds, Brains and Science*. Cambridge: Harvard University Press, 1984.

Shainberg, Lawrence. *Memories of Amnesia: A Novel*. New York: Ivy/Ballantine, 1988.

Shepherd, Gordon M. *Neurobiology*. 2nd ed. Oxford: Oxford University Press, 1988.

Simon, Herbert A. *The Sciences of the Artificial*. 2nd ed. Cambridge: MIT Press, 1981.

Singh, Jagjit. *Great Ideas in Information Theory, Language and Cybernetics*. New York: Dover, 1966.

Spence, Jonathan D. *The Memory Palace of Matteo Ricci*. New York: Viking, 1984.

Von Neumann, John. *The Computer and the Brain*. New Haven: Yale University Press, 1958.

Yates, Frances A. *The Art of Memory*. London: Penguin, 1969.

INDEX

accumulators, 140, 141
acetylcholine, 33
action potential, 11, 90
  firing of, 49, 84
adenosine triphosphate (ATP), 66
adenylate cyclase, 68, 72, 73
adrenaline, 66–7
agonists and antagonists, 82, 85
AIDS research, 114
algorithms, 130, 134, 146, 151, 161, 163, 175
Alkon, Daniel, 74–6, 85, 89, 90
Amaral, David, 88
amnesia, calpain mechanism and, 54–5
amnesiacs, studies of, 35, 55
  R.B., 88, 90, 91
analog net, 155
Anderson, James, 114, 143, 151–3, 161–3, 165, 166, 176, 181, 190
  linear associator, 154–7, 160
animal psychology, 27
Anscombe, Elizabeth, 214
anthropic principle, 113, 210
Aplysia (sea snail), learning in, 23–4, 59, 61, 64–5, *illus.* 69, 153
  and classical conditioning, 70–1, 74
  sensitization with cyclic AMP, 68–9
APV (aminophosphonovaleric acid), 82, 84, 87, 171
artificial brains, 98–9, 134
  see also computer models of the brain; learning machines
artificial intelligence (A.I.), 101, 131, 138, 143, 146–51, 173–6
  companies, 173, 175, 189
  expert system shells, 175, 189
  functionalism and, 223
  neural networks and, 163, 176–7, 179, 182, 186–9, 223
  programs, 147–8, 161, 174–5, 177, 183
artificial neurons, 98
  networks of, 99–100, 128, 134, 151, 154–5
  in perceptron, 140, 141
  see also memory landscape network;

olfactory cortex network; visual cortex network
astronomy, 220–1
atoms
  Bohr model of, 201
  magnetic behavior of, 125–6
  mu-meson, 118
ATP (adenosine triphosphate), 66
Austin, J. L., 214
axon and axon terminal, 20, 48
axon bundle, olfactory, 104

back propagation, 178, 183, 187
Bandelier National Monument, 230–3
Bardeen, John, 111, 117–18, 120, 122
Baudry, Michel, 41–2, 44, 45, 54, 55, 57, 78–9, 86
Bear, Mark, 171–2
behaviorism, 62–3, 98
Bernstein, Jeremy, 138
Bienenstock, Elie, 168
biology, and neural nets, 172, 199
Bliss, Timothy, 36, 89, 90
blood, clotting of, 43
Bobrow, Daniel, 147, 148
Bohr, Niels, 201
Boltzmann, Ludwig, 181
bosons, 120
brain
  as computer, 17–18, 20, 136–8, 152, 163–4, 222, 223–4
  graceful degradation of, 137
  parallel processing by, 137, '150
  perception and memory structures of, 164–5
  plasticity in development of, 170–1
  surgical probing of, 16–17
  theories of mind and, 124–5, 173, 216–17, 222
  as Turing machine, 131, 134
  see also artificial brains; computer metaphor of brain-mind; computer models of the brain; cortex; hippocampus; neocortex; olfactory cortex network; septum

memory
  associative, 80, 170
  biochemical infrastructure of, 13
  biochemical level of, 25
  and brain development in infant,
    170-1
  calpain mechanism and, 42, 92
  clotting of blood as parallel to, 43
  in hippocampus, 91, 107
  holistic and localizationist theories of,
    16-17
  logic and, 164
  molecular theories of, 18-19
  neuroscientists' theories of, 59
  NMDA receptors and, 81, 82, 88, 89,
    90
  as pattern of "lit up" neurons, 20-2,
    23, 73
  perception and, 164-5, 219
  philosophy and, 211, 222
  physicists' theory of, 115, 126
  presynaptic and postsynaptic theories
    of, 61, 73, 88-9
  procedural and declarative, 55, 76,
    87
  scientific theories of, 211
  storage site of, 91
  synapses in short-term and long-
    term, 44
  see also engrams; learning
memory circuits, formation of, 22, 23,
  80, illus. 83
memory landscape network, 179-81
  neurons of, 180
memory model (linear associator), 154-
  8, 162
Memory Palace of Matteo Ricci, The, xiii-
  xiv, 233
metals, electron flow in, 119
metaphysics, 210
Metropolitan Museum of Art, 229
millenarian thinkers, 209
mind
  and brain, 124-5, 173, 216-17
  as computer software, 17-18, 223-4
  as formal system, 222
  theories of, 115, 124
Minsky, Marvin, 138-9, 142-8, 178,
  185-8
  Perceptrons, 143-4, 149, 150, 177, 186,
    187
  society of mind, concept of, 187-8
M.I.T., 138, 143, 152, 153, 186
molecular alphabets of memory, 18-19
Morris, Richard, 87, 198, 199-200
Morse, Samuel, 15

Mosko, Sarah, 33-4
Mountcastle, Vernon, 59, 155
mu-meson catalysis, 118

Nass, Menasche, 151, 168
National Institute of Mental Health, 31,
  74
National Science Foundation, 31
Nature, 56, 198
Nauta, Walle, 32-3, 59
neocortex, 32, 91
Nestor Inc., 112, 189, 190, 191
NETtalk, 182-4, 187
neural nets and neural network models,
  99-100, 134-6, 138, 142-3, illus.
    145, 151, 163, 166
  analog, 155
  based on biology, 172, 199
  computer simulations, 163, 176-7
  as engineering problem-solving, 199-
    202
  engrams in, 157, 184
  learning machines, 135, 138-43, 176
  linear associator, 154-8, 160, 162
  with nonlinear neurons, 162
  of olfactory cortex, 102-7, 189
  with probabilistic neurons, 162
  as surface in multidimensional space,
    179-80
  vectors in, 155-6, 158-61, 169, 179-
    80
  of visual cortex in cats, 166-70
  see also neural network field; neural
    network theory
neural network field, 128, 143, 149, 161-
  2, 193
  artificial intelligence and, 150-1, 163,
    177, 179, 182, 186-8, 223
  companies, 189-90
  conferences, 182, 185
  interdisciplinary interest in, 177-8
  interest of scientists in, 181-2
  neuroscience and, 177, 182, 197-9,
    222
  revival, 176-8, 185
neural network theory, 7
  and epistemology, 211
  pioneers of, 112, 114, 132, 177
  Rosenblatt's theorem, 142
  see also memory landscape network;
    visual development, theory of
neural spike, 50-1, 64, 65, 70
neuromodulators, 67
neurons, 19-20, illus. 21
  action potential of, 11
  as analog devices, 155

ABOUT THE AUTHOR

GEORGE JOHNSON is an editor of "The Week in Review" section of *The New York Times* and the author of *Machinery of the Mind: Inside the New Science of Artificial Intelligence*. His work has appeared in *The New York Times Magazine, The New York Times Book Review,* and *The Sciences*. He lives in New York City.